BENEATH OUR FEET

How much do you know about the rocks beneath your feet – the materials of Planet Earth?

How are rocks and minerals created?

How are rocks pushed up to form mountains?

What types of minerals are formed deep within Earth from crushing pressures and scorching temperatures?

What minerals form the meteorites that crash into our planet from space?

Beneath Our Feet: The Rocks of Planet Earth answers all these questions and many more. Ron Vernon reveals the astounding variety and beauty of the rocks and minerals all around us. Especially when viewed through the microscope, rocks show in exquisite detail their makeup from crystals of different minerals. Many stunning images show rocks from deep within Earth that have flowed and deformed, rocks created and altered by heating and melting, rocks ejected from volcanoes, rocks shaped by erosion at Earth's surface, and extraterrestrial rocks that have crashed into Earth throughout its history and are still doing so today. Over 170 spectacular photographs are accompanied by clear and nontechnical explanations of the main processes that are responsible for creating rocks and minerals inside and on the surface of the planet.

Beneath Our Feet is a richly illustrated introduction to the rocks and minerals of Planet Earth for everyone interested in natural history.

Ron Vernon was born in Sydney, Australia, and has lectured at various universities around the world for the past thirty-nine years. He is now Emeritus Professor of Geology at Macquarie University and Conjoint Professor of Geology at the University of Newcastle. He has worked on rocks in Australia, the United States, Scotland, Ireland, and Mexico, and is still active in research, especially on metamorphic rocks and granites. He is particularly interested in the structure of rocks, as seen in the microscope. He is the author of *Metamorphic Processes* (Murby, 1976), is an editor of the *Journal of Metamorphic Geology,* and has been editor of the *Journal of the Geological Society of Australia.* He likes music (classical and jazz), painting, reading, bushwalking and red wine.

Famous full view of Earth, showing Africa and Saudi Arabia, taken by the crew of
the spacecraft Apollo 17 on their way from Earth to the Moon on 7 December, 1972.
The flow patterns of clouds are clearly visible.

Photo from National Space Data Center for Satellite Information,
Principal Investigator, Dr. Frederick J. Doyle.
Space Science Data Center, Goddard Space Flight Center,
Maryland, United States.

Beneath Our Feet
The Rocks of
Planet Earth

Ron Vernon

Emeritus Professor of Geology
Macquarie University
Sydney, Australia

CAMBRIDGE
UNIVERSITY PRESS

PUBLISHED BY THE PRESS SYNDICATE OF THE UNIVERSITY OF CAMBRIDGE
The Pitt Building, Trumpington Street, Cambridge, United Kingdom

CAMBRIDGE UNIVERSITY PRESS
The Edinburgh Building, Cambridge, CB2 2RU, UK   http://www.cup.cam.ac.uk
40 West 20th Street, New York, NY 10011–4211, USA   http://www.cup.org
10 Stamford Road, Oakleigh, Melbourne 3166, Australia
Ruiz de Alarcón 13, 28014 Madrid, Spain

First published 2000

Printed in the United States of America

Typeface Minion 11/16 pt.   *System* Quark XPress^R [GH]

A catalogue record for this book is available from the British Library.

Library of Congress Cataloguing-in-Publication Data

Vernon, R. H. (Ronald Holden)
Beneath our feet : the rocks of planet Earth / Ron Vernon.
p.   cm.
ISBN 0-521-79030-1
1. Rocks 2. Minerals. I. Title.
QE431.2.V47 2000
552 – dc21
00-023666

ISBN   0 521 79030 1   hardback

Contents

Acknowledgments

I thank the following people for assistance in obtaining specimens, diagrams, photographs and technical information: Mr. Alan Bayless (Aber Resources Limited), Professor Myron Best (Brigham Young University), Dr. Alex Bevan (Western Australian Museum), Dr. Ray Binns (CSIRO Division of Exploration and Mining), Mr. Tom Bradley (Macquarie University), Dr. Eric Christiansen (Brigham Young University), Dr. Geoff Clarke (University of Sydney), Dr. Julie Donnelly-Nolan (United States Geological Survey), Ms. Marie Dowling (National Space Science Data Center, Goddard Space Flight Center, Maryland, United States), Dr. Mike Etheridge (SRK Australasia), Mr. Paul Farquharson (Macquarie University), Dr. Dick Flood (Macquarie University), Mr. Andrew Geggie (Geomatics Canada), Dr. Andrew Glikson (Australian Geological Survey Organization), Professor Trevor Green (Macquarie University), Professor Bill Griffin (Macquarie University), Ms. Karen Horan (United States National Geophysical Data Center), Professor Peter Hudleston (University of Minnesota), Dr. Scott Johnson (Macquarie University), Mr. Nigel Kelly (University of Sydney), Dr. John Lusk (Macquarie University), Associate Professor Ruth Mawson (Macquarie University), Mr. Calvin Nicholls (Nicholls Design Inc.), Dr. Geoff Nichols (Macquarie University), Associate Professor Robin Offler (University of Newcastle), Professor Suzanne O'Reilly (Macquarie University), Dr. Scott Paterson (University of Southern California), Mr. Ross Pogson (Australian Museum, Sydney), Mr. Peter Radford (Kevron Aerial Surveys, Pty. Ltd.), Dr. Jim Ryan (University of New Brunswick), Professor Rick Sibson (University of Otago), Dr. Lin Sutherland (Australian Museum, Sydney), Professor John Talent (Macquarie University), Mr. Graham Teale (Teale and Associates), Dr. Roberto Weinberg (Oxford University), Mr. David Wieprecht (United States Geological Survey), Professor Paul Williams (University of New Brunswick), Mrs. Bonnie Wilson (Newcastle), Dr. Chris Wilson (University of Melbourne) and Mr. Pat Wilson (Newcastle).

I am also grateful to the professional geologists and non-geological people, who kindly read all or part of the book and made helpful comments, namely: Ms. Kirstin Arnhold, Professor Mike Brown, Dr. Dick Flood, Dr. Andrew Glikson, Dr. Scott Johnson, Associate Professor Ruth Mawson, Dr. Andrew McCaig, Ms. Amaranta Ruiz-Nelson, Professor John Talent, and Dr. Marcus Tate, as well as my wife Kay Vernon and stepdaughter Olivia Conolly.

The line drawings were made by Dean Oliver Graphics, Pty. Ltd., Earlwood (Sydney), Australia. All photographs are by the author, unless otherwise indicated.

Introduction

W HAT on Earth are you standing on? We are all fascinated by the planets in our solar system, but how many of us know much about the rocks under our feet – the rocks of Planet Earth?

The 1997 Pathfinder expedition to Mars fired the imagination of people all over the world. Millions of viewers watched television reports and visited the NASA web site to learn more about the fragments of rock on the Martian surface. More than twenty years before, many people had shown great interest in rocks on the surface of the Moon. Wonderful as these missions were, it's unfortunate that not many of us have as much interest in the rocks of our own Planet Earth. Generally it seems to be: "Seen one rock, seen 'em all!" Yet, Earth rocks are far more beautiful and variable than rocks of the Moon and Mars. That's what this book is all about – to show you how beautiful and variable Earth rocks really are.

What's the best way to look at rocks? Unfortunately, most of the rocks we see are in dull grey, weathered, lichen-covered outcrops on hillsides or in fields (Figure 1). So it's worth stopping at road cuttings, quarries, and washed river exposures, where rocks are generally much cleaner. Regions scraped clean by glaciers are also great for looking at fresh rock (Figure 2). A hammer helps a lot because breaking open rock samples reveals the fresh material just inside the weathered rind (Figure 3), but be careful and make sure your eyes are shielded from flying rock chips! It's best to use a geological hammer, but don't use the sharp end except for prying fossils out of relatively soft rocks. Another good way to see the internal structures of rocks is to have a close look at the polished stone facings of bench tops or city buildings. Don't worry if people stare at you. Maybe they'll become interested too!

But rocks are even more spectacular when sliced very thinly and viewed in a microscope. The front cover photo is a beautiful example of Earth rocks seen in the microscope. So is Figure 4, which shows almost exactly the same kind of rock as the broken sample in Figure 3.

Figure 1.

Outcrops of granite, south of Cooma, southeastern
Australia. The outcrops have been rounded by
reactions of the rocks with the atmosphere
("weathering"). The dull grey, weathered surface,
covered with lichen, makes it difficult to appreciate the
beauty of the fresh rock just below the surface.

You'll notice how different the rock looks, and how you can see everything much more clearly in the microscope. This is the way to really see what Earth rocks are made of! In fact, the microscope opens up a whole new world of wonder and beauty, as illustrated by the many microscope photos of rocks in this book.

As well as showing you many colour photos, this book will describe, in the most general and nontechnical way, how Planet Earth behaves, so you can get at least a broad idea of how different rocks form. You won't need an understanding of chemical formulae, and I have tried to keep explanations as simple as possible. I should emphasize that this is *not* meant to

Figure 2.
These magnificent mountains of fresh granite have been exposed by the melting of a glacier in Yosemite Valley, California, United States. When the glacier was active, the moving ice scraped clean the rock surface and protected it from reaction with the atmosphere.

Figure 3.

Block of fresh granite knocked off an outcrop with a geological hammer, showing crystals of different minerals, some light, others dark in colour. The coin is 2 centimetres in diameter.

be an introductory textbook on geology. It's just meant to give you enough information to appreciate the very basic processes responsible for the beautiful rocks of Planet Earth.

Even if you concentrate on the pictures, rather than the text, you will still be able to gain an appreciation of the extraordinary beauty and variety of the materials produced by our wonderful planet. However, if you'd like a little more background information about the photos, the following chapter will explain briefly how microscope slides of rocks are made and what the colours in the photos mean.

Figure 4.

This is what granite like that shown in Figure 3 looks like in the microscope. The rock has been sliced and ground down to make a "thin section" (see Chapter 1). The slice is so thin and the magnification so great that individual crystals are clearly visible. This photo shows how much more detail you can see in the microscope. The colours are not the true colours of the minerals, but are "interference colours," as explained in Chapter 1. The granite is from Dunkeld in the Bathurst area, west of Sydney, New South Wales, Australia. The base of the photo is 1.8 centimetres long.

Looking at rocks and minerals

What are rocks and minerals?

A *mineral* is a naturally occurring inorganic chemical compound, and a *rock* is a solid aggregate of minerals. Minerals and rocks are given names, and you'll encounter a few of these in this book, but not too many! A glossary of some common minerals and rocks is given at the end of the book, but you don't need to refer to it unless you are really interested, as I have tried to keep technical words to a minimum.

Minerals are *crystalline solids,* which means that their atoms are bound tightly together in regular patterns. This is in contrast to *liquids,* in which atoms and small groups of atoms are constantly bonding together and breaking apart. The regular arrangement of atoms in crystalline solids is responsible for the crystal shapes of the spectacular mineral specimens we see in museums (Figure 5).

However, a few natural rocks are not crystalline, for example, **volcanic glass** (Figure 6). We are all familiar with glass and use it every day. Glass is obviously a solid, which means that its atoms are bound tightly together. However, in contrast to crystals, the atoms in glass are arranged irregularly, as they would be in a liquid. The reason glass is non-crystalline is discussed in Chapter 6.

If minerals have regular shapes (for example, if they grow freely in a liquid), they are called **crystals** (Figure 5), but if they have irregular shapes (for example, if their growth is

Figure 5.
Crystals of amethyst (purple quartz) that grew in a water-filled cavity
in solid rock. The crystals grew from the edges of the cavity towards
the centre. The crystals in many museum specimens grew in this way.
The pocket knife is 9 centimetres long.

impeded by other minerals) or if they are fragments, they are called **grains.** Nevertheless, the
internal atomic structure of a particular mineral remains the same, regardless of whether it is
present as crystals or grains in a rock.

How can we learn more about rocks and minerals?

We can learn a lot by viewing rocks on Earth's surface, provided they are unweathered or
"fresh" (Figure 2). Therefore, geologists break samples off rock outcrops with a hammer in

Figure 6.
Volcanic glass, with the curved fracture pattern typical
of broken glass. The specimen is 16 centimetres across.

order to inspect fresh surfaces of the rock (Figure 3). Cylinders of rock ("drill core"),
extracted by drilling down into rocks with a rotating diamond bit, also provide fresh samples
of rock (Figure 7). This is especially useful for geologists exploring for economic mineral
resources underground. Examination of rock samples with a magnifying glass shows us even
more about the minerals present and their relationships to each other (that is, the rock's
structure).

However, we can learn much more with a microscope. The rock is sliced with a diamond-
impregnated circular saw (diamond being the hardest mineral, it cuts all the others), after
which the slice is stuck to a glass microscope slide with a strong transparent glue. Then we
grind down the slice on rotating wheels with abrasive powder (generally carborundum)
until it is only 0.03 mm thick, at which stage it is known as a "thin section." When you hold a
thin section up to the light or when you shine light through it in a microscope, you can see

that when most common minerals are as thin as this, all or most of the light passes through them. Minerals that transmit all the light are colourless, whereas those that transmit some wavelengths of light and absorb other wavelengths are coloured.

When the thin section is examined in a microscope (Figure 4), the individual crystals are magnified, which helps us identify them. Also, because the rock slice is so thin, individual crystals are seen alongside their neighbours, rather than being piled confusingly on top of other crystals, as they are when you look at ordinary rock samples (Figure 3), even with a magnifying glass. This enables us to examine the structure of the rock in detail.

Furthermore, a microscope with Polaroid filters (as used in some sunglasses) enables us to shine polarized light through the minerals. Minerals interfere with the polarized light very differently, depending on their crystal structure. The resulting effects (called "interference

Figure 7.
Cylinder ("drill core") of fresh rock, about 1,700 million years old, extracted by drilling into rocks of the Jervois area, central Australia. The piece of core is 17 centimetres long.

colours") can be very spectacular, as we will see in many photos in this book (for example, the front cover photo and Figure 4). Interference colours provide us with extra ways to identify minerals and reveal their mutual relationships.

In this book, photographs of thin sections taken with these polarized light effects are labelled "interference colours" (implying that the colours you see are not the true colours of the minerals), whereas those taken without polarized light effects are labelled "ordinary colours" (implying true colours).

That's all the background information you need to start enjoying the rock photographs. We will now look at the various ways in which Planet Earth behaves and how this is reflected in the diversity of Earth rocks.

Juggling plates

The rocks beneath our feet

The rocks on which we live provide us with most of our essential materials, such as building materials, road aggregate, sand, clay, concrete, glass, metals, ceramics, gemstones, coal, petroleum products, artesian water, and soil for plants. They are the exposed part of Earth's *crust,* which is a very thin "skin" on the planet. The crust is so thin, relative to the rest of the planet, that it appears only as a line in Figure 8, which illustrates the main "shells" of rock inside Planet Earth.

In fact, Earth is essentially a huge mass of hot, slowly and continually moving rock, which dominates and determines what happens to the thin crust. Therefore, a simple idea of Earth's interior helps us to understand processes that have produced and modified the rocks of the crust that we see exposed at Earth's surface.

Earth's interior

Earth is a series of "shells" (Figure 8). The centre of Earth consists of a spherical *core,* which is composed mainly of the metal *iron,* with some nickel. The central part of the core is solid and the outer part is liquid.

The core is surrounded by the *mantle,* which consists of solid rock rich in oxygen, sili-

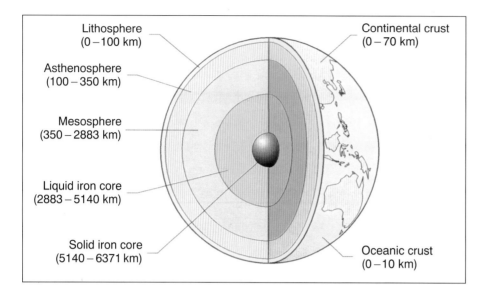

Figure 8.

Sketch showing the main shells of Earth's interior. The crust is so thin that it cannot be shown accurately on a sphere as small as this.

con, and magnesium, with many other chemical elements in smaller amounts. The mantle is not as dense as the core and consists of three shells (Figure 8). In the *asthenosphere* ("weak sphere") the rocks are weaker than in the *lithosphere,* which is a relatively thin zone, extending from the surface down to about 100 kilometres in the mantle. The importance of these differences in rock strength is emphasized in Chapter 3.

The *crust* surrounding the mantle is cooler, lighter, and more rigid than the mantle, and is quite variable in thickness (Figure 9). The thickness of *oceanic crust* averages 6–8 kilometres, whereas the thickness of *continental crust* averages around 30 kilometres and varies from about 20 to about 70 kilometres (Figure 9). Oceanic crust consists mainly of volcanic rock with an average density of about 3.0 grams per cubic centimetre, whereas continental crust consists of a wide variety of rocks with an average density of about 2.8 grams per cubic centimetre.

How do we know that these different compositional shells really exist inside Earth? The information is provided by earthquakes (Chapter 6), which send shock waves through the planet. These earthquake waves travel with speeds that depend on the density of the rocks through which they pass. This means that changes in density cause changes in speed of

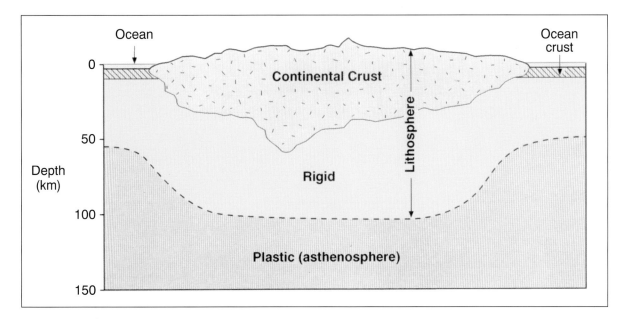

Figure 9.
Sketch showing the difference in thickness between oceanic and
continental crust. It also shows the rigid lithosphere (including the
crust) resting on the much weaker (plastic) asthenosphere.

waves, and so we can detect changes from one type of rock to another. By measuring the
times of arrival of earthquake waves in various places on Earth's surface, knowing the place
where they originated, these changes in Earth's internal structure can be worked out.

Chemical composition of Earth's crust

It may be surprising to learn that the most abundant chemical element in Earth's crust is
oxygen (47 percent by weight and 94 percent by volume!), followed by silicon (28 percent by
weight, but only 1 percent by volume) and aluminium (8 percent by weight, but only 0.5 percent by volume). In fact, oxygen atoms are so much bigger than most of the other atoms in
common minerals that the crust is really oxygen atoms packed closely together, with the little
silicon and aluminium atoms (together with smaller numbers of atoms of other elements,
such as iron, calcium, sodium, potassium, and magnesium) tucked in between.

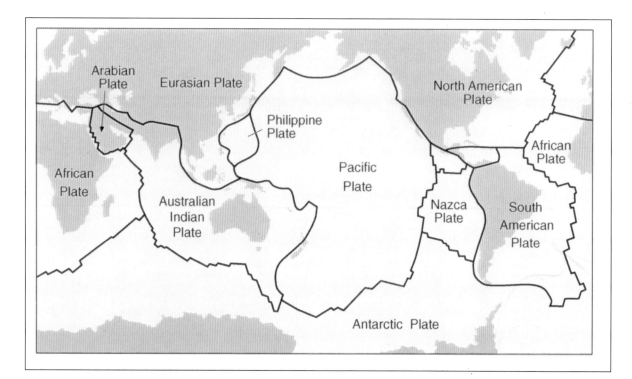

Figure 10.

Sketch showing the distribution of the main lithospheric plates on Earth's surface. Note that the plates have both oceanic (blue) and continental (pink) crust on their surfaces.

Therefore, most common minerals are chemical compounds of oxygen, silicon, and aluminium, with smaller amounts of other elements. These compounds are known as "silicates" and "aluminosilicates," depending on whether aluminium is present or not.

Movement of Earth's lithospheric plates – the actions of a restless planet

The crust and outermost mantle make up a relatively rigid layer called the *lithosphere* (Figures 8, 9), which is divided into a dozen or so *plates* or large fragments (Figure 10). These "lithospheric plates" are thousands of kilometres across and are separated by large fractures. The thickness of the plates varies from only a few kilometres along ocean ridges (Figure 11) to over 100 kilometres beneath some large continents (Figure 9).

The plates are constantly moving very slowly – breaking up, separating, colliding and rubbing past one another – rather like plates of sea ice in polar regions. The relatively rigid plates can move because they rest on the much less rigid asthenosphere (Figure 9). What forces cause these movements?

Convection in Earth's mantle

Heat is transferred upwards from the core towards the surface, partly by rising melted rock (Chapter 5), but mainly by bodily movement of solid rock. In other words, the solid asthenosphere *flows* – very slowly, of course. Why is the asthenosphere so weak it can flow? Part of the reason is that it is locally molten, but even the much more abundant, unmelted, crystalline part is weak enough to flow. This is because it is very hot. The concept of *flow of solid rock* may seem strange at first, but Chapter 3 may help to explain it.

Hot rock tends to rise because it is lighter than cooler rock of the same composition. Movement of hot solid rock and melted rock through Earth's mantle sets up gigantic "convection currents" in the relatively weak – though still mostly solid – mantle rock. These slow convection currents carry the slowly flowing hot rock upwards, spread it out sideways, and then carry it down again as it cools and so becomes denser (Figure 11), just as convection currents operate in heated water. The moving mantle drags along the base of the lithosphere and so causes plate movements (Figure 11).

In other words, Earth's internal "heat engine" – involving continual movement of hot rock – is responsible for the distribution of plates on Earth's surface (Figure 10) and the breakup, movement, and constant reshuffling of continents and oceans.

Why is Earth so different from the Moon?

In contrast to Earth, the Moon has no mountain ranges or volcanoes younger than about 2,000 million years old. This is because it no longer has enough internal heat to drive convection. It is a static ("dead") planetary body, and its oldest rocks in the *terrae* (the light, rugged highlands) have not changed since they were formed 4,600 million years ago – about

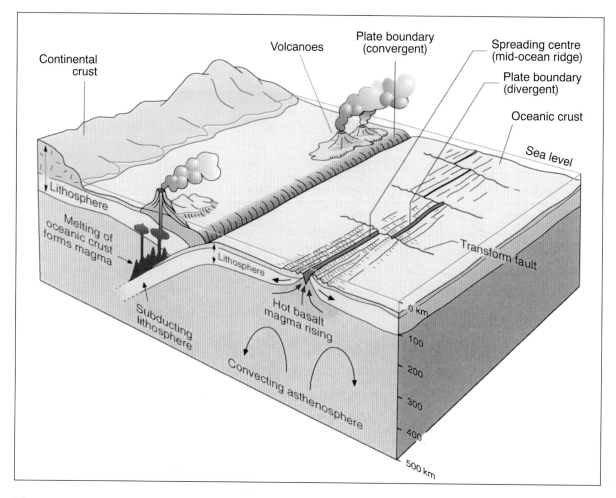

Figure 11.

Sketch showing the main features of mid-ocean ridges and
subduction zones. It also shows (in a very approximate way) the idea
of convection cells in Earth's mantle driving movement of the
lithosphere. Note that melting of the subducted oceanic crust
typically occurs, forming magma that rises and cools to form igneous
rocks, either as intrusions (red) that cool inside the crust or as
volcanoes on the surface (see Chapter 5).

the same time as Earth originated. Rocks formed in the Lunar *maria* (the dark, low basins) have not changed since about 3,200 million years ago.

In contrast, Earth is a dynamic (constantly moving) planet, and this internal movement governs the continual change and recycling of its materials. Its large variety of rocks reflects this slow but relentless dynamic activity over the past 4,600 million years.

Plate boundaries

The boundaries between lithospheric plates are of the following three types (Figure 12).

Divergent (spreading) margins form where two plates move away from each other and melted rock is added from below to each plate as they move apart. Many of these boundaries appear as mid-ocean ridges (Figure 11); this process is referred to as "sea-floor spreading." Divergent margins initiate within continents (e.g., the East African rift zone) and oceans may develop as the two pieces of continental crust separate. However, though it has been suggested that all the oceans began as contintental rift zones, not all of these zones develop into oceans.

Convergent ("subduction") boundaries occur where two plates collide and one slides beneath the other, descending into the mantle, where it is eventually consumed. These boundaries are also known as "subduction zones" (Figure 11). Some convergent boundaries involve collision of continents (Figure 12b), whereas others involve subduction of oceanic crust (Figure 12c).

Passive (transform, shear) boundaries occur as one plate slides past another (Figure 12d), so that the area of the plates is unchanged.

As plates grow at divergent boundaries, they are consumed at convergent boundaries (Figure 11).

Rocks formed at plate boundaries are especially variable and spectacular. Because plate boundaries occur where molten rock can most easily rise from the mantle and the deeper parts of the crust (Figure 11), volcanoes and earthquakes are especially abundant at plate boundaries. Convergent plate boundaries are where most of the deformation of Earth's crust occurs. For example, if two continents collide at convergent boundaries, some of the rocks are pushed upwards, forming Earth's great mountain ranges, such as the European Alps, the Himalayas and the Southern Alps of New Zealand (Figures 13, 14).

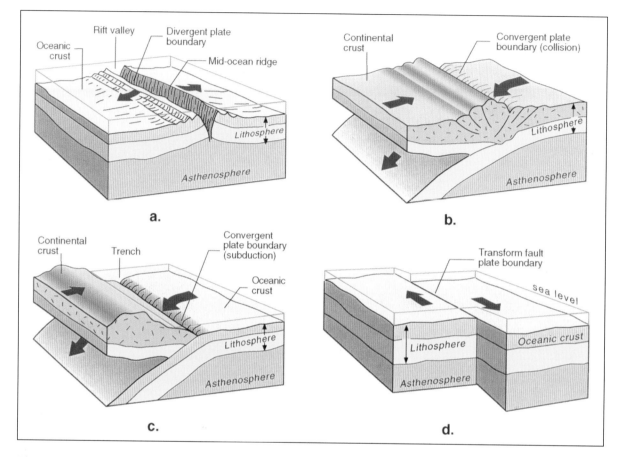

Figure 12.

Sketch showing the three main types of plate boundary, namely divergent (a), convergent (b, c), and passive (d). Note the two main varieties of convergent plate boundary, one (b) involving continental collision and mountain building (see Figure 13), the other (c) involving subduction of oceanic crust.

Figure 13.

Sketch showing the result of collision between continents at a convergent plate boundary. Large folds and fractures ("faults") are formed, as the huge masses of rock are forced outwards and upwards into great mountain ranges, such as the Alps. The pink area with crosses represents an old block of continental crust, onto which the rocks of the other continental block (light brown) have been pushed. From a drawing by Mike Etheridge.

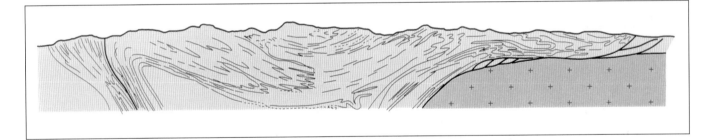

Figure 14.

Mountains made up of layered sedimentary rocks that were deposited 100 million years ago and uplifted about 40 million years ago in the Swiss Alps. The layers were folded and tilted as a result of the collision of the continents of Europe and Africa (Chapter 2). The tilting is shown by the steep attitude of the beds, which controls the slopes of the mountains.

Anything flows

What is flow?

As we have seen in Chapter 2, Earth's mantle is in constant slow movement, despite the fact that most of it is solid. In other words, it *flows* continually, as does much of Earth's crust. At first, you may find the notion that solid materials can flow a very strange idea indeed! But it is extremely important to grasp this concept in order to understand just why Earth is such a dynamic planet.

By "flow" we mean that a material changes its shape in response to a force without breaking. In contrast, "fracture" results when a material cannot flow, but breaks instead.

We are familiar with the idea of fluids (liquids and gases) flowing. For example, Earth's atmosphere flows (changes its shape) in response to the forces of atmospheric currents, and water in a stream flows in response to the force of gravity. This is because fluids have little strength, owing to the weakness of bonds between groups of atoms in the gas or liquid. The atoms can easily slip past each other as the fluid changes its shape.

It's also easy to understand how molten materials flow, for example, metal in a foundry and lava from a volcano (Figures 15, 16). These are formerly solid materials that are liquefied (melted) by heating. They flow because their large amount of heat energy enables their atoms and groups of atoms to break apart and remain separate long enough to move past each other as the material changes its shape – for example, as lava flows downhill in response

Figure 15.

Red-hot molten rock (lava) flowing underneath a cooled crust of
solid volcanic rock (basalt), Kilauea Volcano, Hawaii.

Photo by Suzanne O'Reilly.

to the force of gravity. When these melts cool, their atoms become bound more tightly
together, and so they flow more slowly. Eventually they stop flowing and freeze into a solid
material, in which the atoms are permanently bound together.

However, it is less obvious that *all solid materials can flow in the right circumstances.* Yet
we can see a lot of evidence for this in rocks exposed on Earth's surface. For example, compare the patterns formed by foam transported by the flow of water in a mountain stream
(Figure 17) with the almost identical patterns formed by the flow of solid rock deep in
Earth's crust (Figure 18). How can solid rocks flow like this?

Figure 16.
Microscope photo showing strong alignment of feldspar (orthoclase) crystals formed by flow of lava. The long crystals were rotated until they were almost parallel to each other during movement of the lava, and were preserved in that orientation as the lava solidified. Interference colours; base of photo 4.4 millimetres.

How can solids flow?

Most of the evidence that solid rocks flow comes from rocks formed in the deeper parts of the crust. These rocks have been uplifted by Earth movements and exposed at the surface by erosion. About 15 kilometres or more below the surface, the rocks are so hot that they flow, rather than break, when they are compressed and sheared very slowly by plate motion (Chapter 2).

As the rocks flow, they become distorted into beautiful *folded patterns* that can be seen on all scales – on the sides of mountains (Figures 19, 20, 21), in outcrops (Figures 22–25), in rock specimens (Figure 26), and in thin sections viewed in the microscope (Figure 27). Folded patterns in glaciers (Figures 28, 29) indicate that ice can also flow on Earth's surface.

These regular folded patterns form because most of the mineral grains and crystals in the folded rocks change their shapes by *flow,* without breaking. Individual grains may become very elongated or may become converted to flowing, stretched out aggregates of much smaller new grains (Figure 30). Minerals with their atoms arranged in sheet-like structures, such as mica, generally bend and kink (Figure 31).

The fact that rocks and minerals remain *solid* as they flow indicates that bonds between atoms can be broken during flow without causing the minerals to develop cracks. How does this take place? The answer is basically simple in outline, though more complicated in detail.

(text continues on page 41)

Figure 17.
Complex fold patterns formed by flow in water, outlined by foam on
the surface of a mountain stream.

Figure 18.
Complex fold patterns formed by flow in a solid rock that was very strongly deformed when it was hot, deep in Earth's crust, approximately 1,100 million years ago in Ontario, southeastern Canada. The folded patterns are very similar to those in Figure 17. Geological hammer for scale.

Figure 19.

Rocks folded into spectacular patterns, about
1,000 million years ago, exposed in a cliff in
the Oygarden Islands, East Antarctica, with
glacier ice in the foreground. Person and sled
for scale.

Photo by Nigel Kelly.

Figure 20.

Very complicated folded patterns formed by intense squeezing when these rocks were buried deep in Earth's crust about 1,000 million years ago, Oygarden Islands, East Antarctica; glacier ice in the foreground. Photo by Nigel Kelly.

Figure 21.

Folded rocks (about 1,000 million years old) on the side of Mount
Lanyon in the northern Prince Charles Mountains, Antarctica, with
glacier ice in the foreground. Also present are light greenish-grey
zones with dark green fragments, which cut the folds and so were
formed later. These are brittle fault zones containing "breccia" (see
Chapter 6). The photo illustrates the difference between structures
formed by flow (folds) and structures formed by fracture (faults).
Photo by Geoff Nichols.

Figure 22.

Sedimentary rocks with contrasting dark- and light-coloured beds
(Chapter 7) that were heated and folded about 500 million years ago,
Kangaroo Island, South Australia. The coin is 2.9 centimetres in
diameter.

Figure 23.

Sedimentary rocks that have been heated and
intricately folded, Harcuvar Mountains, Arizona,
United States. The pen (centre) is 15 centimetres long.

Figure 24.

Strongly folded rocks in the Swiss Alps. The light-coloured layer to the left of the hammer shows an early fold that has been re-folded, indicating that two deformation events affected these rocks when they were buried in Earth's crust.

Photo by Scott Paterson.

Figure 25.

Very severe squeezing and folding of solid rock, about 1,100 million
years old, in Ontario, southeastern Canada. The flow of the rock has
been so intense that a formerly straight, continuous dark layer has
been squeezed into very tight folds and separated into isolated lenses.
The knife (centre) is 9 centimetres long.

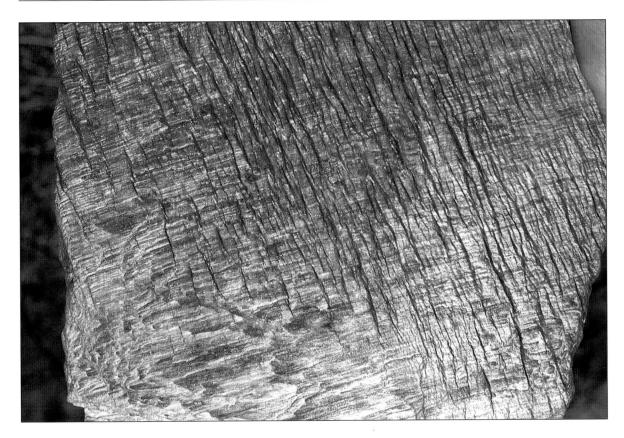

Figure 26.

In this rock (about 400 million years old), an intense foliation formed by one episode of strong deformation was folded ("crenulated") by a second episode of deformation. All the deformation was accomplished by flow of the solid rock. Abercrombie River, southeastern Australia. The block of rock is about 30 centimetres across.

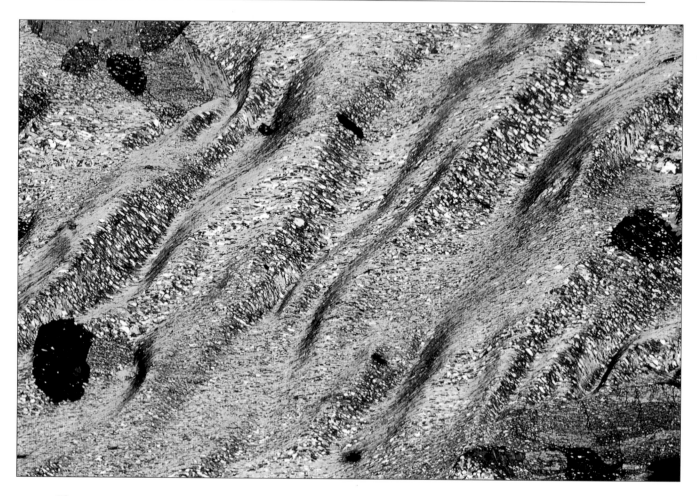

Figure 27.

Microscope photo showing small folds ("crenulations"), similar to
those shown in Figure 26, formed by flow in a rock that was heated to
about 550° C, Picuris Range, New Mexico, United States. Also shown
are large crystals of garnet (black) and staurolite (yellow, at bottom-
right of photo); these minerals are mentioned in Chapter 8.
Interference colours; base of photo 1.8 centimetres.

Figure 28.

Tightly folded layers in glacial ice, exposed in an ice cliff, 15 metres high, at the northwest shore of Generator Lake, Baffin Island, Northwest Territories, Canada. The dark and light layering is caused by variations in the number of air bubbles and/or the amount of dirt accumulated in the ice.

Photo by Peter Hudleston.

Figure 29.

Aerial view, looking vertically downwards, of the end of an alpine valley glacier (the
Crusoe Glacier) to the northwest of the Barnes Ice Cap, on Axel Heiberg Island,
Northwest Territories, Canada, showing large folds caused by the flowing of solid ice
on Earth's surface. The folds are outlined by (1) thick, dark layers of rock debris
("moraine") carried along on the surface of the glacier and (2) numerous thinner
layers of "dirty" ice all through the glacier. Crevasses (fractures) formed at the surface
of the flowing glacier are also present. The base of the photo is 2.8 kilometres long.
Photo from the National Air Photo Library, Geomatics Canada.

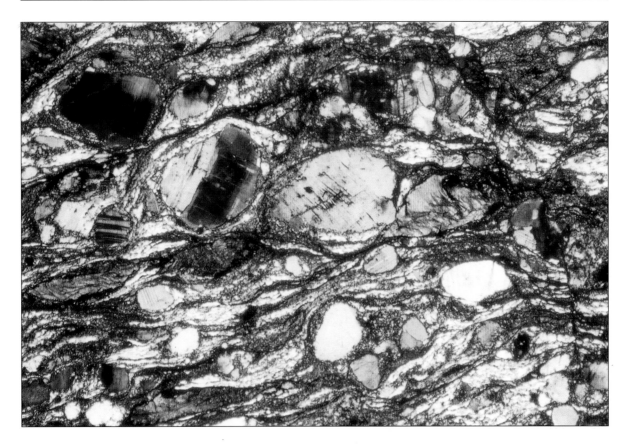

Figure 30.

Microscope photo of a deformed granite, about 250 million years old, from
Wongwibinda, New South Wales, Australia. Large, strong grains of feldspar (shades
of grey) show evidence of internal bending and breaking. In contrast to this
behaviour, biotite mica (brown) has been stretched out by flow into elongated lenses,
and quartz (white to grey) has been drawn out into very thin ribbons and converted
("recrystallized") into aggregates of very small grains in the process. This rock would
have started out resembling the granite of Figure 3, which shows just how powerful
Earth forces can be. Interference colours; base of photo 4.4 millimetres.

Figure 31.
Microscope photo of crystals of mica (muscovite) that have been bent and kinked by flow during deformation when the rock was buried about 12 kilometres deep in Earth's crust about 500 million years ago; Mount Lofty Ranges, South Australia. Interference colours; base of photo 4.4 millimetres.

However, it's worth taking a little time to think about, as it explains so much about how movement occurs in Earth's solid crust and mantle – movement that accounts for the distribution of Earth's continents and oceans.

A good analogy for understanding this movement is to imagine moving a large carpet on a floor. If you want to move it, you can take hold of one of the ends and try pulling it. However, this requires a lot of force, owing to the weight of the carpet and the frictional forces that cause it to stick to the floor. A much easier way to move it is to form a hump or bulge parallel to the edge of the carpet and simply move the hump along (Figure 32). This only requires overcoming the friction underneath the hump, not the whole carpet. As you move the hump along, the carpet settles back onto the floor behind you and pulls away from the floor under the new position of the hump. Eventually the hump comes to the end of the carpet, which will then have moved a small distance, equivalent to the size of the hump. If you make and move many of these humps parallel to each other, you will eventually move the carpet to where you want it.

Figure 32.

Sketch illustrating how a carpet can be moved along a floor by forming and moving a hump. This is much easier than trying to move the whole carpet at once by tugging at one end because only the frictional forces under the hump need to be overcome at any one time. Comparing this with Figure 33, you can see that the floor is analogous to the slip plane in a crystal and the hump is analogous to the dislocation.

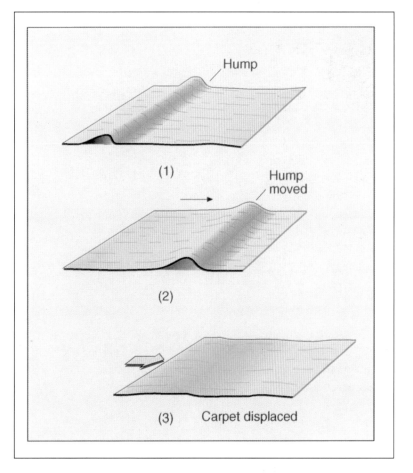

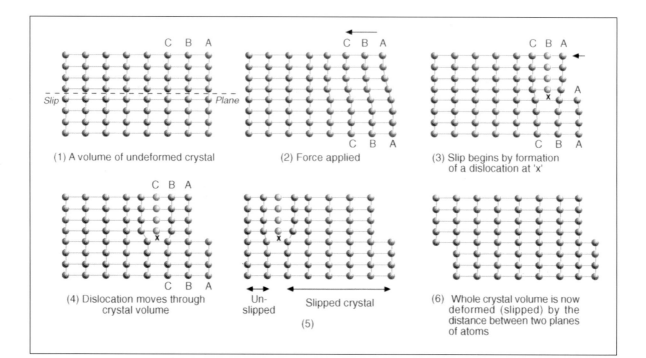

(1) A volume of undeformed crystal

(2) Force applied

(3) Slip begins by formation of a dislocation at 'x'

(4) Dislocation moves through crystal volume

(5) Un-slipped / Slipped crystal

(6) Whole crystal volume is now deformed (slipped) by the distance between two planes of atoms

Figure 33.

Sketch illustrating how a "dislocation" moves through part of a mineral, which is represented by a simplified grid of atoms, with bonds between them represented by lines; in real minerals, the atoms would touch each other, but they are separated in the sketch to show each atom clearly. In stage 1 the crystal is undeformed, but it has a potential plane of atoms (a "slip plane") along which deformation can occur if a force is applied. The force begins to be applied in stage 2. By stage 3, the atoms in plane A-A above the slip plane have linked up with the row of atoms B-B below the slip plane, forcing half of the B-B plane of atoms to break away along the line X (coming straight at you from the page), which is called a *dislocation*. This extra half plane of atoms then links to the C-C row of atoms below the slip plane, forcing half of the C-C plane of atoms to break away along the line X (stage 4). In this way, the dislocation X (with its extra half plane of atoms) can move right through the volume of crystal, producing a very small permanent deformation (stage 6). The dislocation marks the boundary between deformed ("slipped") and undeformed crystal, as shown for stage 5. Note that this process is not the same as going directly from stage 1 to stage 6, which would involve forming a fracture, requiring breaking of *all* the bonds along the slip plane at the same time.

Modified from Hobbs, Means & Williams (1976)
"An Outline of Structural Geology," (Wiley, New York), fig. 2.2, p. 81.

Now comes the more technical explanation, which you can skip if you feel inclined! A flowing mineral does something broadly similar to the hump in the carpet. The force on the mineral grain causes *one* row of atoms at a time to break. Then the next row breaks and the one behind it joins together again. Therefore, successive rows break one at a time until the break ("dislocation") moves right through the mineral, causing a displacement of one row of atoms. If many thousands of these minute displacements occur, they can cause the mineral to change its visible shape. Each little break needs only a very small amount of energy, and the process does not require the mineral to change the overall arrangement of its atoms. Figure 33 explains the process, if you would like to follow it a little more closely. It is completely different from deformation by fracture, which involves the breaking of many rows of atoms at the same time.

The important point to remember is that solid crystalline minerals can change their shapes without breaking, in response to an applied force, using minute amounts of energy repeatedly, rather than by using a lot of energy all at once.

A final point to note is that when rocks are being deformed while hot, the flowing mineral grains typically convert to new, smaller grains as the flow proceeds – a process called "recrystallization." This process reduces the buildup of strain in the mineral by providing new, undeformed grains that can continue to deform, thus allowing the rock to continue to change its shape by flow, rather than by breaking. Many strongly deformed rocks contain deformed old grains that have been partly replaced by new (recrystallized) grains (Figures 34, 35, 36). Generally these new grains are polygonal in shape, but they may be elongate in minerals with sheet-like atomic structures, such as mica (Figure 37).

When do rocks flow and when do they break?

If we hit pieces of rocks from Earth's crust with a hammer, they break. But if we squeeze them very slowly under high confining pressure (for example, in a tight metal cylinder or under high gas pressure) they flow – especially if they're hot.

Experiments have shown that the main factors favouring flow of solid crystalline rocks are: (1) high confining pressure (which makes it difficult for the rock to expand and hence break during deformation), (2) high temperature (which allows dislocations to move freely through minerals), and (3) slow application of the deforming force (which gives the dislocations enough time to move). Therefore, we should expect flow to dominate in the deeper parts of Earth's crust and in the mantle, where the rocks are hot and under high confining pressures.

(text continues on page 46)

Figure 34.

Microscope photo showing a large, strongly deformed, old grain of calcite (calcium carbonate) that has been partly replaced during deformation by new (recrystallized) grains of calcite, in a rock from the Mount Lofty Ranges, South Australia. The rock was deformed about 500 million years ago at a depth of about 15 kilometres and a temperature of around 400° C. This thin section is only 0.007 millimetres thick, which shows the calcite grains more clearly than in a normal thin section (0.03 mm thick). Interference colours; base of photo 3.5 millimetres.

Figure 35.
Microscope photo showing large, bent, strongly deformed grains of plagioclase that have been partly replaced by smaller, new (recrystallized) grains of plagioclase during deformation at a temperature of over 500° C, in a 1,100 million-year-old rock from northwest South Australia. Interference colours; base of photo 4.4 millimetres.

Figure 36.

Microscope photo showing large, strongly deformed grains of olivine, locally
recrystallized to aggregates of much smaller polygonal grains of olivine, in a peridotite
deformed in Earth's mantle. The rock was brought to the surface as a fragment in
basalt lava (see Chapter 4). Interference colours; base of photo 2.5 centimetres.

Flow in Earth's outer mantle (asthenosphere)

A good example of hot flowing rock is **peridotite** (olivine-rich rock) in Earth's outer mantle
(Chapter 2). The large amount of heat energy enables some of the atoms in the olivine to
break their bonds, so that dislocations move around relatively easily (though the rock
remains solid and crystalline). This movement of atoms through the solid olivine grains
helps them to change their shapes and to convert to new grains (Figure 36).

Figure 37.
Microscope photo of a crystal of mica (biotite) that has partly
recrystallized to smaller elongate crystals during deformation deep in
Earth's crust, about 1,600 million years ago, Broken Hill, western New
South Wales, Australia. Interference colours; base of photo 3.5
millimetres.

The surprising thing about glacier ice

Another important but perhaps astonishing example of hot flowing rock is glacier *ice*. Ice is
a mineral because it is a naturally occurring, inorganic, crystalline chemical compound
(with the composition of water); we use the same name for a rock formed by aggregates of
this mineral. The surprising thing about glaciers is that ice is even closer to its melting temperature than most rocks of Earth's asthenosphere are to their melting temperatures!

In fact, at low and middle latitudes and relatively low altitudes, ice is close to or at its
melting point, so that ice and water occur together. However, at high altitudes and latitudes,
for example in polar sheet glaciers, the ice is solid right through the ice sheet. Nevertheless, it
is no more than 15–25° C below its melting point of 0° C. Though we generally think of it as

being cold, ice in glaciers in fact is very hot, relative to its melting temperature! Therefore, it's not really surprising that ice *flows* in response to the force of gravity on and just below Earth's surface (Figures 28, 29). Moreover, it recrystallizes to polygonal aggregates of new grains (Figures 38A,B), just as most other common minerals do when deformed at high temperatures (Figures 34–36). Ice flows and recrystallizes at such low temperatures because the

Figure 38.

(**A**) Microscope photo of strongly deformed, natural glacier ice from the same area as shown in Figure 28, consisting of polygonal grains of ice. The specimen was obtained by drilling and extracting a core of the ice, making a slice, and grinding it down until the ice was thin enough to transmit light. Interference colours; the faint grid pattern (used for a scale) represents 1 centimetre squares.

(**B**) Microscope photo showing an aggregate of polygonal grains formed during flow of solid ice in a laboratory experiment designed to simulate flow in natural ice. Interference colours; width of sample 3 centimetres.

Photo by Peter Hudleston.

A

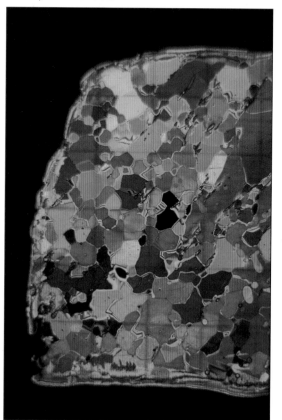

Photo by Chris Wilson.

B

bonding between water molecules in ice is much weaker than the bonding between atoms in common minerals in Earth's crust and mantle.

Another important point is that the internal parts of ice sheets are under a relatively high confining pressure (from the weight of the ice), which helps promote flow. This flow of solid ice inside the glacier, assisted by slip between the ice and the bedrock at the bottom of the glacier, enables the glacier to move steadily downhill at a rate of a few centimetres to a few metres per day.

Where the edges of the ice are confined, as in valley glaciers, the centre flows faster than the edges, forming spectacular folded patterns (outlined by rock debris carried along on the surface of the glacier) (Figure 29). Folded patterns can also be seen in polar sheet glaciers, especially near their edges and in ice sea cliffs (Figure 28). The folds are outlined by layers caused by variations in the amounts of clear ice, ice with air bubbles, and ice with accumulated "dirt" (small mineral particles).

Flow or fracture?

On the other hand, the ice in many glaciers, especially at their tops, is not under high confining pressure from above, and if the glacier has to change its shape abruptly – for example, where it passes over a sharp change in slope – the ice is forced to flow more rapidly. As it cannot deform fast enough by flow, it breaks, forming fractures up to 50 metres deep called *crevasses* (Figure 39). Similarly, in the upper parts of Earth's crust, where temperatures and confining pressure are relatively low, rocks tend to break rather than flow (Figure 40). The fractures are called *faults* (Chapter 6).

The difference between the appearances of faults (Figure 40) and folds (Figures 19–25) reflects the difference between fracture and flow, respectively. Both faults and folds can occur in the same rock outcrop, where they may be related to the same deformation event, for example if the rocks deform by folding until they are forced to break, and then continue to deform by faulting. On the other hand, faults may be formed in a later deformation event. For example, Figure 21 shows folds formed by flow when the rocks were relatively hot, cut by faults formed later when the rocks had cooled down enough to inhibit flow.

The distribution and nature of folds and faults give us a very general idea of the depth at which Earth forces operated at the time these structures were formed, in the particular part of the crust concerned. Generally fracture dominates at depths of less than about 15 kilometres and flow dominates at greater depths in Earth's crust.

Figure 39.

Crevasses in a glacier, formed by fracturing, where the
ice travels over a sharp change in slope from the
snowfield (top) into the main flowing glacier (carrying
dark layers of rock debris) in the valley below. The ice
breaks because it is forced to move faster than it can
achieve by flow. The larger crevasses are filled with
younger (white) snow. Unterer Theodulgletcher,
viewed from the Gornergrat, Swiss Alps.

Figure 40.
Small faults that have displaced layers of different composition, in a
1,600 million-year-old rock at Olary, South Australia. The knife is 9
centimetres long.

4

A mantle of green

What is Earth's mantle made of?

When I was very young, grownups used to tell me that the Moon was made of green cheese. Now that rock samples have been collected from the Moon, we know that it consists of mainly of grey rocks, very like some of the basalts on Earth (Chapter 5). But it may come as a surprise to learn that much of Planet Earth – the mantle – in fact is green!

How do we know that Earth's mantle is green? The answer is that fragments of the mantle are constantly being ripped off and brought to Earth's surface by rapidly rising molten rock (magma), and are ejected from volcanoes. Because they rise so quickly and are cooled so rapidly in the atmosphere, the minerals in the fragments don't have time to change to minerals that would be more stable at the surface. This gives us the opportunity to observe minerals from deep inside Earth.

The mantle fragments are hurled out of volcanoes with some of the lava sticking to them, twisting as they fly through the air, and cooling to form aptly named "volcanic bombs" (Figure 41). When a bomb is broken open or sawn through, the beautiful green fragment of Earth's mantle is revealed inside the dark grey solidified lava (Figure 42).

From these mantle fragments preserved in volcanic rocks, we learn that the main mineral of Earth's outer mantle is **olivine,** which, as the name suggests, is olive-green (Figure 42).

Figure 41.

Volcanic "bomb" thrown out of a volcano (now extinct), Mount Noorat, near Terang, western Victoria, Australia. Hot, fluid lava clung to a fragment of mantle rock (peridotite) brought up in the magma, and hardened around the fragment. The bomb twirled as it flew through the air, causing the lava to solidify into a characteristic ellipsoidal shape. The bomb is 36 centimetres long.

Sample courtesy of John Talent.

Figure 42.

When volcanic bombs (Figure 41) are broken open, many of them reveal fragments of green peridotite from Earth's mantle, such as the one shown here. Note the granular structure of the peridotite, the abundant bright olive-green olivine, and the less abundant, darker green pyroxene. A microscope photo of a similar rock is shown in Figure 43. The specimen is from Mount Leura, an extinct volcano in western Victoria, Australia. The scale is in centimetres.

Photo by Suzanne O'Reilly.

Because the gem variety of olivine is "peridot," the rock of the outer mantle is called "peridotite." **Pyroxene,** the other common mineral in the mantle, is also green (Figure 42). The olivine and pyroxene tend to have polygonal grain shapes, not regular crystal shapes, so that the rock looks granular (Figure 43). This is a common structure in solid rocks that have been heated for a long time, and is typical of many of the rocks in Earth's mantle and deeper crust (Chapter 8).

Another source of information on Earth's outer mantle is provided by large slices of mantle rock that have been caught up in major faults during the movement and collision of continents (Chapter 2). These solid slices of peridotite are squeezed upwards into the crust in situations such as that shown in Figure 13, and thus can be exposed at Earth's surface after deep erosion has occurred. Flowing solid peridotite may also rise in mid-ocean ridges, where new basalt magma is being added to the oceanic crust (Figure 11).

Water percolating through the rocks of the crust typically affects these bodies of mantle peridotite. Because olivine is stable only at high temperatures, the water reacts with it and changes it to **serpentine** (Figure 44), which is a water-rich mineral stable at lower temperatures in the crust. Serpentine is rather easy to deform, and so some altered peridotites show evidence of intense deformation (Figure 45). The deformation of rocks is discussed in more detail in Chapter 10. Other examples of the reactive effects of water, which is so common in Earth's crust, are discussed in Chapter 9.

Figure 43.

Microscope photo of mantle rock (peridotite) ejected as a fragment (similar to the one shown in Figure 42) in a volcanic eruption on Hawaii, showing the bright interference colours of olivine and pyroxene, and the granular structure that is typical of rocks that have been hot for long periods (Chapter 8). Interference colours; base of photo 1.8 centimetres.

Figure 44.
Microscope photo of mantle rock (peridotite from New Caledonia) that has been thrust into Earth's crust and partly reacted with water to form the water rich mineral serpentine. The serpentine (grey colours) has replaced the olivine (bright colours) along cracks and around grain edges. Interference colours; base of photo 3.5 millimetres.

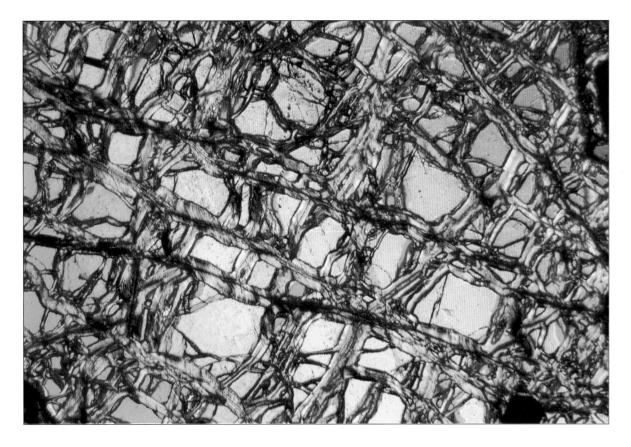

Figure 45.

Microscope photo showing small folds ("crenulations") in a peridotite from Sweden, containing olivine (bright colours) and serpentine (grey-blue). The rock was formed in Earth's outer mantle and later deformed in Earth's crust. Interference colours; base of photo 1.9 millimetres.

5

Hot stuff

Because of their terrifying power and their capacity for devastation, as well as the beautiful symmetrical mountains they build (Figure 46), volcanoes capture everyone's imagination. All volcanic eruptions are spectacular, but some are more destructive than others. All of them produce beautiful rocks with a variety of structures, especially when viewed in the microscope, and we will look at many examples. Where does the molten rock ("lava") that pours out of or explodes from volcanoes come from?

Magma

Molten rock (together with crystals that precipitate from the melt or are broken off solid rocks) is called **magma.** Rocks that form by solidification of magma as it cools are called *igneous rocks.*

In hot magma, atoms and groups of atoms join together and break apart constantly. As the magma cools, the atoms move more slowly, and eventually join together permanently to form minute solid aggregates that grow to form **crystals.** This freezing of magma is known as *crystallization.* It is essentially similar to the freezing (crystallization) of water to ice. Eventually, crystallization becomes so advanced that the magma solidifies.

If magma reaches Earth's surface, it is called **lava,** which freezes rapidly in the cool atmosphere to form *volcanic rock.* These rocks are said to be "fine-grained" because they have very

Figure 46.

Mount St. Helens, Washington, United States, a typical symmetrical volcano built up of layers of lava and volcanic rock fragments deposited over a long period of time, photographed on May 17, 1980 – one day before the famous eruptions that blew out the centre of the mountain.

Photo by Harry Glicken. By courtesy of the United States
Department of the Interior, U.S. Geological Survey,
David A. Johnston Cascades Volcano Observatory,
Vancouver, Washington, United States.

small crystals and grains. Some lava fails to crystallize at all because the cooling is so fast that atoms are not given enough time to organize themselves to form crystals. The result is a solid rock, but without a crystalline structure, called **volcanic glass** (Figure 6). This is very like ordinary domestic glass, which is made by the very rapid cooling of melted silicate minerals.

If magma is trapped below the surface, it cools slowly to form *intrusive rock*, typically with large crystals and grains (Figures 3, 4); such a rock is said to be "coarse-grained." The magma intrudes and pushes aside solid rocks in Earth's crust as it rises. It may squeeze between layers in the solid rocks (Figure 47) or cut across them by forcing open cracks (Figures 48, 49). It may also dislodge and incorporate fragments of the rocks it intrudes (Figure 50). When exposed at Earth's surface, some intrusions of igneous rock are more resistant to weathering than the rocks they intruded, and so are left projecting above the more eroded rocks (Figure 51).

The structures we see in solidified igneous rocks depend a great deal on the *fluidity* of the magma concerned. Some magmas are hot and relatively *fluid,* whereas others are cooler (though still very hot!) and more *viscous* (stiff). We will examine each type in turn.

(text continues on page 66)

Figure 47.
Intrusion of light-coloured granite into dark-coloured, layered, strongly deformed metamorphic rocks (Chapter 8), Depot Peak, MacRobertson Land, Antarctica. This relatively small intrusion pushed its way in between the rock layers before it crystallized and became too viscous to move any further. The rocks are about 1,000 million years old.
Photo by Geoff Nichols.

Figure 48.
Intrusion of dark-coloured basalt into light-coloured
granite, Mount Desert Island, Maine, United States.

Figure 49.
Intrusions of light-coloured, very coarse-grained granite (pegmatite) into dark-coloured metamorphic rock (Chapter 8), near Fort William, northwestern Scotland. The intrusions probably forced open large fractures.

Figure 50.

Intrusion of light grey granite into darker-coloured, layered metamorphic rocks (Chapter 8), Mount Desert Island, Maine, United States. The granite pushed its way between the layers and wedged the rocks apart. Eventually it broke the layers and engulfed them as fragments in the magma.

Figure 51.

Shiprock, New Mexico, United States, which consists of remnants of intrusive igneous rocks that are more resistant to weathering than the rocks they originally intruded. A volcano formerly covered these remnants, but because it consisted mainly of fragmental material it has been eroded away. This weathering exposed the last pulse of magma that cooled and solidified in the channels that fed the volcano.

Photo by Scott Johnson.

Hot, fluid magma

Sometimes disturbances in Earth's mantle cause it to partly melt. Partial melting of the asthenosphere (Chapter 2) at about 100 kilometres depth or less produces **magma** (molten rock plus any unmelted crystals), which rises rapidly because it is lighter than the surrounding unmelted rock. It may enter Earth's crust and often reaches the surface.

The magma produced by this melting of the asthenosphere is the hottest and most fluid of magmas. If it reaches Earth's surface and cools rapidly it is called **basalt** (Figures 52, 53) and if trapped below the surface (i.e., in Earth's upper mantle or crust) and cools slowly it is called **gabbro** (Figures 54, 55). The chemical composition of both rocks is the same; only the grain size is different because of the different cooling rates.

(text continues on page 71)

Figure 52.

This dark grey, fine-grained basalt from west of Lismore, New South Wales, Australia, may appear to be the dullest looking rock in this book! In fact, it looks very much like most of the rocks on the Moon. It solidified from fluid lava on Earth's surface, so that it crystallized rapidly. Also present are scattered larger crystals that grew slowly when the magma was held at depth before eruption. This rock has approximately the same chemical composition as the gabbro shown in Figure 54. The sample is 10 centimetres long. Compare this with the microscope photo of a similar rock shown in Figure 53.

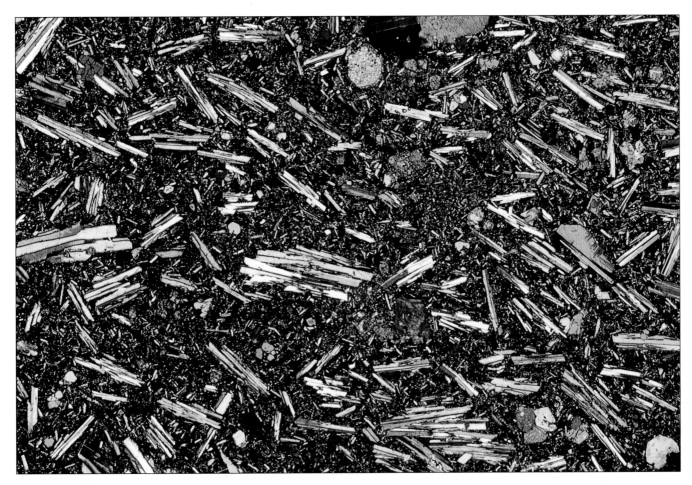

Figure 53.

Microscope photo of basalt, showing crystals with simple regular shapes. The brightly-coloured crystals are olivine and pyroxene, and the grey crystals are plagioclase (a very common mineral in most rocks). The rather dull-looking basalt of Figure 52 would look very much like this in the microscope, which shows how well the microscope can reveal the detail and beauty of rocks. Interference colours; base of photo 1.8 centimetres.

Figure 54.

Coarse-grained gabbro, which solidified from magma trapped below Earth's surface, so that it crystallized slowly. This rock has approximately the same chemical composition as the basalts shown in Figures 52 and 53. The coin is 2.9 centimetres in diameter. Compare this with the microscope photo of a similar rock shown in Figure 55.

Figure 55.

Microscope photo of gabbro from the central Sierra Nevada, California, United States, showing crystals with simple regular shapes, but larger than the crystals in basalt (Figure 53). The brightly coloured crystals are olivine and pyroxene, and the grey crystals are plagioclase. This gabbro is similar to the specimen illustrated in Figure 54. Interference colours; base of photo 4.4 millimetres.

Figure 56.

Mauna Kea, Hawaii, a shield volcano, 4,200 metres high, formed by the accumulation of fluid basalt lava flows produced by relatively quiet eruptions.

Photo by Suzanne O'Reilly.

Because the basalt lava is hot (about 1,100° C) when it is erupted, it is relatively fluid. As a result, it spreads out rapidly over Earth's surface and can cover large areas. The basalt lava fills river valleys and commonly builds broad, flat volcanoes called "shield volcanoes" (Figure 56) or even large plateaus of lava flows piled one on top of the other. The surface of the hot flowing basalt lava cools to form twisted, ropy shapes called "pahoehoe" (Figures 15, 57), but as the flow cools, slows down, and becomes stiffer, the solid crust may break up into rough blocks known as "aa" (Figure 58). The "pahoehoe" and "aa" terms are Hawaiian and have been adopted because many of the early observations on basalt lava flows were made on Hawaii.

Figure 57.
Ropy basalt ("pahoehoe") with a smooth, shiny surface, Mauna Ulu, Hawaii. The cooled, solidified surface of the lava flow shown in Figure 15 has a similar structure. The twisted ropy shapes reflect the high fluidity of the lava when it was moving. Also shown is the basalt mound (miniature shield volcano) around the vent from which the lava flowed, and a forest of trees that were burnt by the heat of the lava.
Photo by Suzanne O'Reilly.

Figure 58.

Clinkery basalt ("aa"), Mauna Ulu, Hawaii. The rough fragmental shapes reflect the
later stage of cooling of the lava, when the surface became more viscous than in the
ropy lava in the background (also see Figure 57). Broken-up ropy lava is present in
the foreground.

Photo by Suzanne O'Reilly.

Basalt magma contains only a small amount of water (up to about 2 percent by weight).
When the lava is erupted onto Earth's surface, this water forms small bubbles (because it is
no longer under high pressure) and escapes easily from the hot, fluid basalt lava. Therefore,
basalt eruptions are generally nonexplosive or only weakly explosive, and commonly produce "spatter cones" and fluid lava flows (Figures 15, 59, 60). Of course, this is only relative, as
the eruptions are still quite spectacular! If the gas escapes as the lava solidifies, the shapes of
the gas bubbles may be preserved in the solid rock (Figure 61). Much later, these cavities may

Figure 59.

Eruptions of basalt lava are relatively quiet because gas can escape easily from the fluid lava and so doesn't build up excessive pressure. This is an aerial view at night of an eruption forming a "lava fountain" at the Pu'u O'o vent on Kilauea Volcano, Hawaii, on April 14, 1986. The red-hot (1,100° C) lava flowed slowly as a series of braided streams down the shallow slopes of the shield volcano (Figure 56).

Photo by Lyn Topinka. Courtesy of the United States Department of the Interior, U.S. Geological Survey, David A. Johnston Cascades Volcano Observatory, Vancouver, Washington, United States.

become filled with new minerals precipitated from water circulating through the solid basalt along cracks (Figures 61, 62).

Because Earth's mantle is rich in magnesium-rich silicates that are stable only at high temperatures (for example, olivine and pyroxene), the magma produced by melting of the mantle is also relatively rich in these chemical compounds, and thus it precipitates olivine and pyroxene as it begins to cool (Figure 53). These minerals are accompanied by the very common mineral, **plagioclase** (an aluminosilicate of calcium and sodium). The crystals grow in the liquid melt and develop regular shapes (crystal forms) if the cooling rate is slow enough (Figure 53).

(text continues on page 77)

Figure 60.

"Spatter" of red-hot basalt lava erupting from a fissure
on Kilauea Volcano, Hawaii, on September 20, 1977.

Photo by United States Geological Survey (Hawaii Volcano
Observatory), courtesy of the National Oceanic and
Atmospheric Administration/National Geophysical
Data Center, Boulder, Colorado, United States.

Figure 61.

Rounded shapes of gas bubbles preserved as holes in solid basalt, Peats Ridge, west of Sydney, New South Wales, Australia. Most of the bubble holes have been coated with blue-green chlorite and a few holes have been almost filled by fibrous white zeolite. These minerals were deposited from hot watery solutions travelling along small cracks over a long period of time. The sample, which is a sawn slab, is 28 centimetres long.

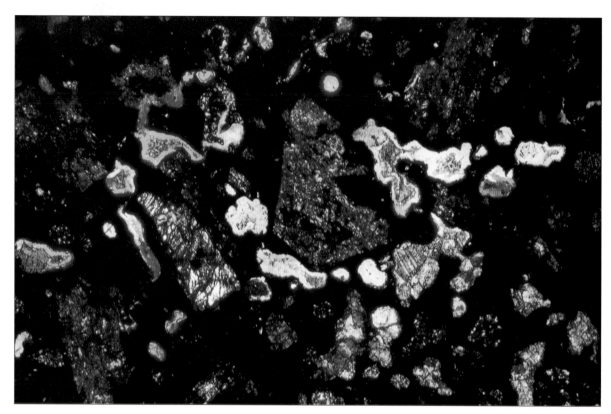

Figure 62.

Microscope photo of a basalt, about 350 million years old, from
Keepit, New South Wales, Australia. The photo shows many gas
bubble holes that have been completely filled by minerals precipitated
from hot watery solutions that percolated through the rock along
cracks for millions of years after it solidified. The mineral with the
bright colours is a water-bearing calcium aluminosilicate (prehnite).
Between the bubble holes, the rock consists of altered glass (black)
and crystals of plagioclase (now completely replaced by other
minerals, especially prehnite, owing to attack from the same solutions
that filled the bubble holes). Interference colours; base of photo 4.4
millimetres.

Crystal skeletons

If basalt lava cools very rapidly (for example, if it is erupted quickly into the sea), the crystal shapes are more intricate – with spiky, branched or skeletal shapes (Figures 63–65). Why do these complicated crystal shapes develop?

Because the rapid cooling slows down the rate at which atoms can move though the liquid melt, the atoms not needed by the growing crystal cannot move away from the crystal surface into the liquid as fast as previously. Therefore, they accumulate around the growing crystal as a layer of "waste" or "impurity" atoms, which stops the crystal from growing in the normal way. To keep the crystal growing, spikes or branches develop on the surface of the crystal, and pass through the polluted layer into the fresh liquid outside (Figure 66). This liquid contains the atoms needed for the crystal to continue growing. Of course, the spike itself

Figure 63.
Microscope photo of basalt from Lake Superior, Canada, showing plagioclase crystals with spiky shapes formed during very rapid cooling of the lava. Interference colours; base of photo 5 millimetres.

Figure 64.

Microscope photo of ancient (about 2,800 million years old) magnesium-rich basalt
from Western Australia, showing an olivine crystal (colourless, at right) with spikes
that have grown out from its corners (see Figure 66), in response to very rapid
cooling of the lava. Also shown are very spiky, branched and fern-like crystals of
pyroxene. Normal colours; base of photo 3.5 millimetres.

develops a layer of polluted liquid around it, and new spikes have to develop. Therefore,
under these conditions, crystals grow by continually branching out into fresh liquid, forming
branched, treelike, fernlike or skeletal shapes.

In some situations of very rapid cooling, the remaining liquid part of the lava may fail to
crystallize at all because of insufficient time for the atoms to arrange themselves into crystals.
The result is **volcanic glass,** occurring between the crystals that have already formed (Figures
67, 68).

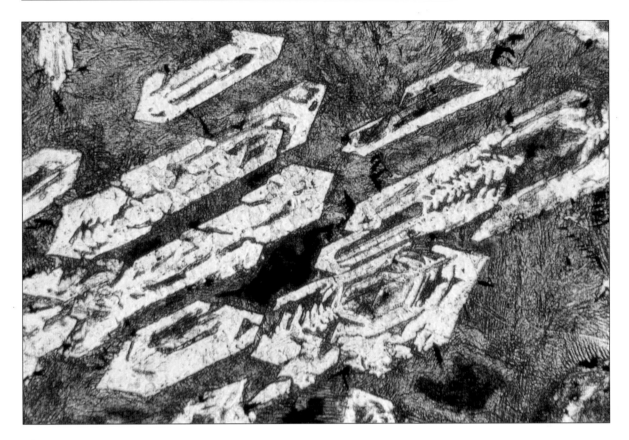

Figure 65.

Microscope photo of ancient (about 2,800 million years old) magnesium-rich basalt from Western Australia, showing olivine crystals with skeleton-like shapes formed in response to very rapid cooling of lava. The olivine crystals are aligned in response to flow of the magma as it solidified. Very spiky, branched, and fern-like crystals of pyroxene are also present. Normal colours; base of photo 3.5 millimetres.

Basalt magma reaches the base of Earth's crust with a temperature of about 1,000 to 1,200° C. Because the crust at these depths is generally at a temperature of only 800–900° C, the basalt magma may heat up the rocks of the crust and partly melt them. This produces a different kind of magma (see next section), which itself begins to rise through the crust. In many places throughout the world, we can see evidence that the two magmas have mixed and mingled in beautiful intricate patterns deep in the crust (Figure 69).

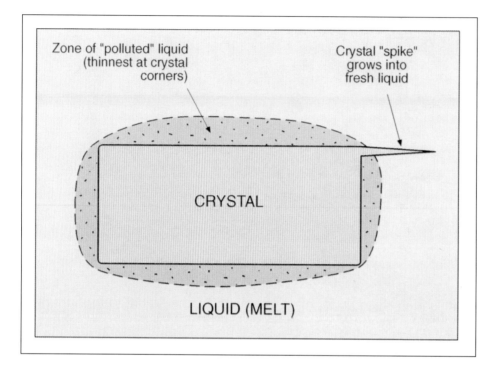

Figure 66.

Sketch to show zone of melt "polluted" by chemical components unwanted by the growing crystal in a rapidly cooled magma. This polluted zone is narrowest at the corners of the growing crystal, because the "impurity" atoms can escape into the liquid most easily at those points. Spikes grow out from the corners because of the shorter distance to "fresh" liquid, in which the crystal can continue to grow. Compare this sketch with the shape of the olivine crystal on the right side of Figure 64.

Figure 67.

Microscope photo of basalt (about 350 million years old) from the Hunter Valley, New South Wales, Australia. The magma initially cooled slowly at depth, forming the large crystals of plagioclase (colourless), pyroxene (light green) and magnetite (black), after which it was erupted onto Earth's surface, where it cooled so rapidly that the rest of the rock formed glass (grey material between the crystals). Normal colours, base of photo 4.4 millimetres.

Figure 68.

Same view as shown in Figure 67, but showing interference colours. The plagioclase is in shades of grey, the pyroxene is more brightly coloured, and the glass is black. (Because glass doesn't have a regular arrangement of its atoms, it doesn't interfere with light as crystals do, and so it doesn't produce interference colours.) Interference colours; base of photo 4.4 millimetres.

Figure 69.
Mingling of viscous granite magma (light coloured) with more fluid gabbro magma
(dark); central Sierra Nevada, California, United States.

Cooler, viscous magma

This new magma produced by melting rocks of Earth's crust is much richer in silicon, aluminium, potassium, and sodium than basalt magma. The igneous rocks formed from this kind of magma are generally lighter in colour. If it reaches Earth's surface and cools rapidly, the magma crystallizes to form **rhyolite.** If the magma is trapped below the surface and cools slowly it forms **granite.**

Because it cools slowly, granite magma has time to grow relatively large crystals (Figures 3, 4). Most granites are relatively even grained, but some have very large crystals ("megacrysts"), as shown in Figure 70. A few consist of huge crystals; these rocks are called **pegmatites** (Figure 71). The feldspar crystals in some pegmatites are large enough to mine!

Many granites show alignment of their crystals in response to forces acting on the magma when it was liquid enough to flow, but viscous enough to preserve the alignment (Figure 72). Rare granites develop strange but spectacular, spherically banded ("orbicular") structures (Figure 73), which are of somewhat controversial origin, but that probably developed as a result of addition of water to the magma.

This rhyolite-granite kind of magma is cooler (generally less than 1,000° C) and more *viscous* (stiffer) than basalt-gabbro magma. In fact, it is about a thousand times more viscous. Therefore, rhyolite lava generally flows for only a relatively short distance (rarely more than a few kilometres) before it gets so stiff that it stops flowing. Its high viscosity also preserves beautiful, intricate flow patterns formed as the flow slows down (Figures 74, 75). Many

Figure 70.

Granite (about 350 million years old) with very large pink crystals (orthoclase), near Wellington, western New South Wales, Australia. The orthoclase crystals tend to be aligned, owing to flow of the magma as it solidified. The base of the specimen is 33 centimetres long.

Figure 71.

Pegmatite (very coarse-grained granite) consisting of pink feldspar (orthoclase),
white, milky quartz, and scattered black crystals of biotite mica; Kymi, Finland.
Camera lens cap for scale.

rhyolite flows are so viscous that they cannot leave the vent at all, and solidify as domes and
spines that plug the neck of the volcano (Figures 76, 77). Later eruptions may rupture this
plug, producing a fragmental rock called *volcanic breccia* (Figure 78). "Breccia" is the Italian
word for a break.

As mentioned above, granite magma cooling slowly at depth grows large, well-formed
crystals, but if the magma is suddenly taken to the surface and erupted, the rest of the
magma cools too quickly for large crystals to form. The result is a rock with large crystals
scattered through a much finer-grained aggregate (Figure 79). Rapid cooling can produce
spikes on existing crystals, as in basalts (Figure 80).

Figure 72.
Granite (about 1,850 million years old) from Cape Colbert, Eyre
Peninsula, South Australia, showing parallel alignment of feldspar
crystals in response to flow of the magma. The alignment was preserved
as the magma solidified. Camera lens cap for scale.

Very rapidly cooled rhyolite lava also fails to crystallize and forms volcanic glass
(Figure 6), because the atoms have insufficient time to arrange themselves into crystals.
This rearrangement of atoms is a much more difficult process in viscous rhyolite lava than
in the more fluid basalt lava, because the atoms are quite strongly bound together in large
groups in the rhyolite melt. Therefore, rhyolite glass is much more common than basalt
glass.

Glass is not as stable as crystals, and so in time it tends to crystallize as delicate radiating
aggregates of fibres that grow in the solid glass, typically on existing crystals (Figures 81, 82).

(text continues on page 95)

Figure 73.

"Orbicular" granite from Finland, showing
spectacular distorted spheres (orbs) of alternating
dark- and light-coloured minerals. The coin is 2.7
centimetres across.

Photo by Geoff Clarke.

Figure 74.
Folded flow patterns in rhyolite on Great Cranberry
Island, Maine, United States. The patterns were caused
by flow of very viscous lava on Earth's surface. The
pocket knife is 9 centimetres long.

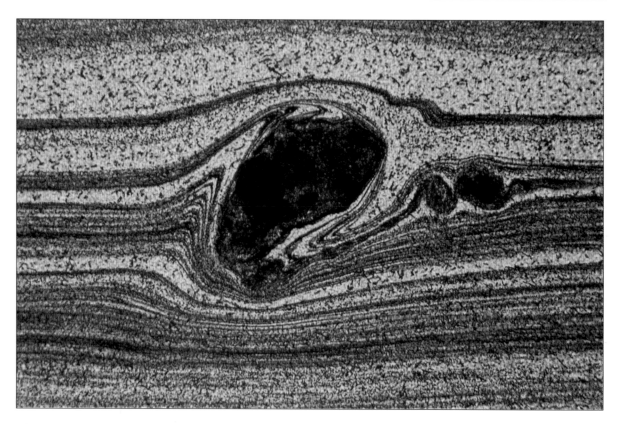

Figure 75.

Microscope photo showing flow patterns in glassy rhyolite from Glass Mountain, Medicine Lake volcano, northeastern California, United States. The patterns were formed when the lava was flowing on Earth's surface, before it quickly solidified into glass. The deflection of the flow lines around the dark solid objects (probably very altered crystal fragments) indicates that the objects were rotated during flow of the lava. The flow lines have been tightly folded around the large solid object during its rotation. Normal colours; base of photo 1.75 millimetres.

Specimen courtesy of Julie Donnelly-Nolan.

Figure 76.

Symmetrical volcano (Mount St. Helens, Washington, United States) after eruptions in 1980 that blew out the centre of the mountain. Later eruptions (for example, the May 1982 eruption shown here) began to build it up again, and a small lava dome inside the main crater can be seen (also see Figure 77). The dome is composed of lava so viscous that it solidified without flowing far.

Photo by Lyn Topinka. Courtesy of the United States Department of the Interior,
U.S. Geological Survey, David A. Johnston Cascades Volcano Observatory,
Vancouver, Washington, United States.

Figure 77.

Lava dome inside the crater of Mount St. Helens,
Washington, United States, in 1984 (compare with
Figure 76), formed by the extrusion of viscous lava
that solidified rapidly around the vent.

Photo by Lyn Topinka. Courtesy of the United States
Department of the Interior, U.S. Geological Survey,
David A. Johnston Cascades Volcano Observatory,
Vancouver, Washington, United States.

Figure 78.

Volcanic breccia, about 350 million years old, from the Hunter Valley, New South Wales, Australia, consisting of angular fragments of rhyolite (with straight and folded flow patterns) disrupted and dispersed by a later explosive eruption that produced an aggregate of small fragments. The knife is 9 centimetres long.

Figure 79.

Sawn slab of rhyolite with large, well-formed crystals
of quartz (dark grey) and feldspar (mainly pink
orthoclase), which grew when the magma was at
depth in Earth's crust. These crystals are dispersed in a
fine-grained aggregate of grains (too small to see with
the unaided eye) formed when the lava was erupted.
The coin is 2.4 centimetres in diameter.

Figure 80.

Many delicate spikes grew on this feldspar crystal
when the lava cooled rapidly. The large feldspar crystal
had grown previously, while the magma was at depth
in Earth's crust. An explanation of the spikes is given
in Figure 66. The rock is about 350 million years old
and occurs near Bathurst, New South Wales, Australia.
Interference colours; base of photo 4.4 millimetres.

Figure 81.

Microscope photo showing radiating, fibrous minerals
that have grown in solid volcanic glass around existing
feldspar crystals, in the Taupo volcanic area, New
Zealand. Normal colours; base of photo 4.4
millimetres.

Figure 82.
Microscope photo showing radiating feldspar fibres that have grown in volcanic glass. The former glass has been completely replaced by the fibrous crystals and more granular aggregates. Interference colours; base of photo 1.3 millimetres.

Explosive eruptions

Though rhyolite lava can and does form small flows, it is generally so stiff and full of gas that it explodes from the vent of the volcano in spectacular and destructive eruptions. During these eruptions, the lava solidifies and shatters into *fragments,* which are scattered over large areas. This explosive kind of eruption is in marked contrast to the typically quieter eruptions of basalt lava, which generally produce flows, rather than fragmental rocks.

The main reason for the explosive eruptions is that the rhyolite-granite type of magma can dissolve more water (up to 10 percent) than basalt-gabbro magma. As the magma rises, cools, and begins to crystallize minerals (most of which don't contain water), the concentration of water in the liquid increases and begins to form gas (superheated steam). The gas builds up pressure inside the magma trapped just below volcanoes until the covering rocks fracture. The release of pressure allows the steam to escape suddenly. This causes an explosive eruption, commonly with devastating effects on the surrounding landscape. A recent example is the eruption of Mount St. Helens (Figure 83).

In fact, major eruptions of rhyolite lava on Earth's surface are nearly always explosive, and the fragmental deposits produced by these huge explosions can be very extensive, as in the western United States. They occur as sheets ranging from a few metres to a hundred metres thick, covering areas of up to thousands of square kilometres. They form such an important type of volcanic deposit and their structures are so distinctive that many photographs of them are included in this book (Figures 84–93).

During an explosive eruption, the escaping gas boils off, leaving the shapes of many small bubbles in the viscous lava, which cools very rapidly (freezes) to form glass. The resulting "glass froth" is called **pumice** (Figures 84, 85). Pumice is very light, owing to the abundance of the air-filled cavities, and may float on water. In fact, the pumice fragments washed up on beaches were formed in explosive volcanic eruptions, perhaps thousands of kilometres away. As in basalt, the bubble holes may remain empty (Figures 84, 85), but when hot watery solutions percolate along small cracks in the pumice for thousands or even millions of years, the holes may become filled with minerals precipitated from the solutions.

During an explosive eruption, the gas continues to escape so fast that it breaks the pumice into large lumps and small, sharp glass pieces ("shards"). It also shatters crystals in the pumice that had crystallized from the magma when it was trapped below the surface and cooling slowly. The high gas pressure forces the mixture of glass fragments, pumice fragments, crystal fragments, and gas high into the atmosphere (Figure 83).

The glass and crystal fragments are mainly of sand size and are called volcanic "ash." Of course, they are not ash in the usual sense of materials left after burning something, but are

(text continues on page 100)

Figure 83. *(opposite page)*
Explosive eruption of Mount St. Helens, Washington, United States, on May 18, 1980. The eruption was characterized by a huge cloud filled with gas and solid fragments (pumice and volcanic "ash"). During the eruption, 400 metres of the peak collapsed or blew outwards.

Photo by Austin Post. Courtesy of the United States Department of the Interior, U.S. Geological Survey, David A. Johnston Cascades Volcano Observatory, Vancouver, Washington, United States.

Figure 84.

Fragment of pumice from the Bishop Tuff, California,
United States, also shown in Figures 89 and 90. The
pumice shows abundant gas bubble shapes. The coin
is 2.4 centimetres in diameter.

Figure 85.

Microscope photo of pumice, showing numerous gas bubble shapes. The bubble shapes in the bottom half of the photo are circular, but those at the top are elliptical because they were stretched out during flow of the viscous magma after the gas escaped. Specimen found on a beach at Herron Island, Australia, by Tom Bradley, who also made the thin section – an extremely difficult task for such delicate material! Normal colours; base of photo 1.75 millimetres.

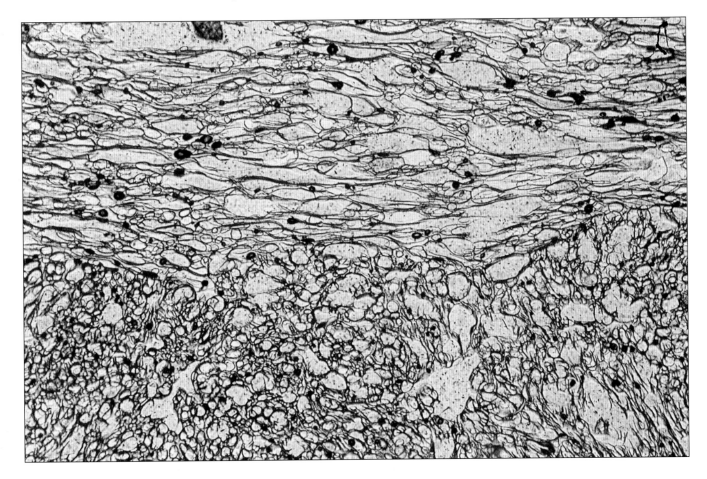

rock and mineral fragments. The falling ash forms a fragmental deposit called *ash-fall tuff*, in which the sharp-cornered, curved shapes of glass fragments can be seen in the microscope (Figures 86, 87). Some of the curved edges of these fragments may be the typical curved fractures that glass produces when it breaks (Figure 6), but most of them probably represent the edges of bubble shapes in the shattered pumice. In fact, microscope photos of ash-fall tuffs show all stages from relatively large pumice fragments, some of which have been stretched during the eruption (Figure 86), through small chips in which only one or two bubble shapes are present, to individual shards (Figure 87).

The high gas pressure may also blow out the side of the volcano, forcing the very hot

Figure 86.

Microscope photo of ash-fall tuff (about 350 million years old) from the Hunter Valley, New South Wales, Australia, formed of fragments ejected in an explosive eruption similar to that shown in Figure 83. The photo shows a fragment of volcanic glass with flow lines and numerous gas bubble holes (almost enough to be called "pumice"), some of which have been stretched by the flow of the viscous lava. Normal colours; base of photo 4.4 millimetres.

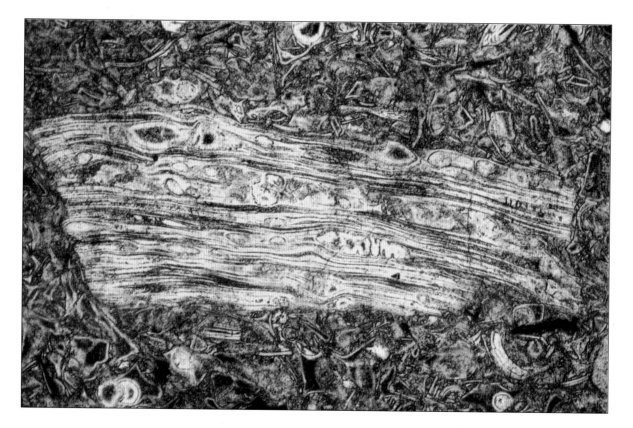

mixture of glass fragments, pumice fragments, crystal fragments, and gas to rush down the slopes at speeds of over 100 kilometres per hour (Figure 88). Such a flow can cause great destruction, as it can spread for up to 100 kilometres. It is a fragmental flow, not a lava flow, and is called an *ash flow*. When the flow slows down, it deposits the ash as a thick deposit (Figure 89), in which the pumice fragments are often clearly visible (Figure 90). The resulting rock is called an *ash-flow tuff*.

Figure 87.

Microscope photo of the same ash-fall tuff as in Figure 86. Apart from a few crystal fragments (for example, the clear quartz fragment at the left side), the fragments are mainly of glass formed by the fragmentation of pumice (similar to that shown in Figure 84). The glass has been replaced by red iron oxide (hematite) and other minerals precipitated from hot watery solutions passing through the rock for a long time after the eruption. A few small fragments of pumice are present, each with one or two complete bubble holes; these holes were filled much later by younger minerals deposited from water percolating through cracks in the rock. Many of the glass fragments ("shards") have sharp corners and curved edges, suggesting that they are remnants of bubble holes in the former pumice. Normal colours; base of photo 4.4 millimetres.

Figure 88.

Ash flow descending the flanks of Mount St. Helens (Washington, United States) on May 18, 1980, at a speed of over 100 kilometres per hour. The flow stretches from the crater to the valley floor below, and is marked by a cloud of steam and volcanic ash.

Photo by Peter Lipman. Courtesy of the United States Department of the Interior, U.S. Geological Survey, David A. Johnston Cascades Volcano Observatory, Vancouver, Washington, United States.

Figure 89.

Thick sequence of tuff units, near Bishop, California, United States – the result of gigantic explosive eruptions about 700,000 years ago. The hat for scale is worn by John Reid.

Figure 90.

Pumice fragments in the tuff shown in the bottom part of Figure 89. A close-up photo of one of the pumice fragments is shown in Figure 84. The knife is 9 centimetres long.

Figure 91.
Specimen of welded tuff (about 400 million years
old) from the Shoalhaven River, New South Wales,
Australia. The dark pumice fragments were
flattened and rounded by heat before they
completely solidified. The specimen is 14.5
centimetres long.

Many ash flows are so hot when they come to rest that the still soft glass and pumice
fragments become compressed and squashed together (Figure 91). The flattened pumice
fragments have distinctive lens shapes and are called "fiamme," which is the Italian word for
"flames" (the singular being "fiamma"). If the glass and pumice fragments become fused
together by the heat and pressure, the rock is appropriately called a *welded tuff*. The formerly
sharp glass fragments become rounded and squashed in the welding process (Figure 92).

Massive explosive eruptions of this kind may extract so much magma (tens to hundreds
of cubic kilometres) from just below the surface that the ground collapses into the empty

Figure 92.
Microscope photo of ash-flow tuff (about 350 million years old) from Barraba, New
South Wales, Australia, showing crystal fragments and curved, angular glass
fragments ("shards") with rounded corners, indicating partial welding. Normal
colours; base of photo 3.5 millimetres.

space, forming an enormous circular depression called a *caldera* (Figure 93). Volcanic craters
and calderas must be carefully distinguished from meteorite impact craters (Chapter 11).

Judging from the widespread, thick deposits of ash-flow tuff and ash-fall tuff (Figures
89, 90) in many places on Earth's surface (for example, the western United States and the
North Island of New Zealand), gigantic explosive eruptions on a scale probably not yet
observed by humans have occurred in the past. They will almost certainly occur again!

Figure 93.

A large circular depression (caldera) formed by a gigantic explosive volcanic eruption 6,600 years ago and now filled with water. This is Crater Lake, Oregon, United States, 8 kilometres in diameter, which occurs at the summit of a former large volcano, Mount Mazama. The island (Wizard Island) is a small cone formed by a minor ejection of fragmental volcanic material after the immense eruptions and consequent collapse of the volcano that formed the caldera.

Photo by Scott Johnson.

Diamonds!

Most diamonds crystallize deep in Earth's mantle, though some of the more recently discovered diamonds appear to have been formed in the deep crust. The diamonds are brought up towards the surface by magma.

The popularity of diamonds as gems is due not only to their brilliance and comparative rarity, but also to their extreme hardness – they don't scratch when worn as a ring, for example. It's hard to imagine a greater contrast than the hard brightness of diamond and the soft dullness of graphite (used in pencils). Both are minerals and both have the same chemical composition – they consist simply of carbon. However, they have very different arrangements of their carbon atoms. In diamond, the carbon atoms are arranged in a strong, three-dimensional network, whereas in graphite the carbon atoms are arranged in sheets, between which the bonding is very weak; this enables the sheets to slide on one another, which makes graphite a good lubricant and enables the dark colour to slide off the pencil tip onto the paper.

Why does diamond occur in some rocks and graphite in others? The answer lies in the temperature and pressure conditions at which the rocks are formed. Diamonds are formed in places inside Earth where the ratio of pressure to temperature is high. They are brought towards the surface very rapidly in magma that erupts explosively, owing to carbon dioxide gas and steam that escapes from the magma as it rises. The result is a fragmental igneous rock called "kimberlite" (Figure 94), because of its original discovery at Kimberley in South Africa. Fortunately, the rise and eruption of the magma is so fast that the diamonds don't have enough time to transform to graphite, which is the more stable form of carbon at and

Figure 94.
Fragmental igneous rock ("kimberlite") from deep in Earth's mantle.
This kind of rock may carry diamonds – an example is shown in
Figure 95. Fragments of rocks from Earth's mantle and crust are very
common in kimberlites. The sample is 18 centimetres across at its base.

close to Earth's surface. Moreover, the bonds between the carbon atoms in diamond are so strong that it doesn't decompose to graphite, even when exposed to the atmosphere for long periods – in natural rocks or in jewellery.

We are used to seeing cut and polished diamonds in jewellery, but diamonds can also be beautiful in their original uncut condition in natural rocks, as shown in Figure 95.

Figure 95.

Beautiful crystal of diamond (the Aber Diamond) in "kimberlite" from deep in Earth's mantle, discovered in drill core at the Diavik Project, 300 kilometres northeast of Yellowknife, Northwest Territories, Canada. The diamond is about 8 millimetres across and weighs 1.75 carats. This is how diamonds occur in nature, before being extracted from the rock, cut, and polished. This rock also contains crystals of red-brown garnet (top of photo).

Photograph by Calvin Nicholls. Courtesy of Nicholls Design Inc.,
Lindsay, Ontario, Canada and Aber Resources Limited, Vancouver, Canada.

Breaking point

If rocks are deformed quickly, especially at low temperature and low confining pressure, they tend to break, forming *faults*. They also break if they are forced to deform beyond the limit that flow can accommodate the applied forces, as when glacier ice flows over a cliff (Figure 39).

Once a fault is formed and the Earth forces acting on it exceed a certain critical value, the blocks on either side of the fault move relative to each other. The results of some of these fault movements are well known to us as *earthquakes*. If the fault is relatively high in Earth's crust, the walls of the fault grind against each other and form fragmental rocks along the fault.

Faults are most common in the outer 15 kilometres of Earth's crust because the rocks in the outer crust are relatively cool and under less confining pressure than the rocks deeper down. Moreover, *water* is more abundant in the outer crust and this may help rocks to break by infiltrating between crystals. It may also lubricate fault surfaces. However, faults can occur at any depth in Earth's crust, provided the flow strength of the rocks is exceeded.

Where rocks break and form small openings, water enters from the adjacent rocks because the pressure in the openings is less than that in the wall-rock. The local reduction in pressure may also cause minerals, especially quartz, which were dissolved in the water, to precipitate as veins (Figure 96). Gold may also be deposited in quartz veins of this type.

Water travelling along major faults may encounter an opening formed by irregularities

Figure 96.
Veins of quartz (white) formed in fractures that opened during brittle deformation
of the rock, Towamba River, southeastern New South Wales, Australia. The knife is
9 centimetres long.

in the fault surface. For example, a step in the fault may occur (Figure 97). Because such an
opening is a site of lower pressure than the surrounding rock pressure, the wall-rocks may
expand into the cavity and break into fragments, filling the opening with loosely packed
pieces of wall-rock. The wall-rocks are said to "implode" into the cavity. Water travelling
along the fault may deposit minerals between the fragments, forming a solid rock called a
"fault breccia" (Figures 98, 99). "Breccia" is the Italian word for a break.

Glass in faults

If Earth forces cause movement on a fault deep in the crust, where the rocks are dry (so that
no water is available to lubricate the fault) and the confining pressure is high (forcing the

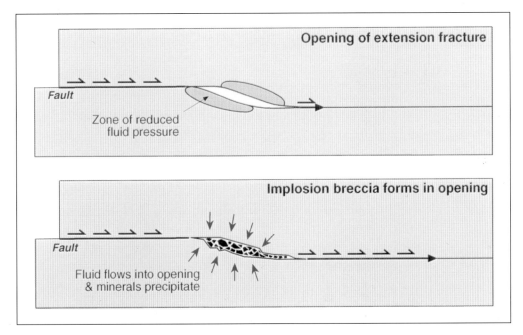

Figure 97.

Diagram showing formation of an opening in a step in a fault (or a join between two faults) resulting in implosion of the wall-rocks into the opening to form a fault breccia. Redrawn from a diagram by Mike Etheridge.

walls of the fault tightly together), frictional resistance to movement may be so high that the rocks along the fault rapidly build up heat and begin to *melt*. The melting occurs for only a very short time and then the melt cools very rapidly to form dark-coloured rocks resembling volcanic glass (Figure 100). During this rapid cooling, crystallization of the melt may occur as spiky and feathery outgrowths (with shapes similar to those discussed in Chapter 5) on existing crystal fragments (Figure 101). However, if the cooling is too fast for much crystallization of the melt to occur, it solidifies as *glass* (Figure 102). In time, the glass itself may crystallize as radiating aggregates of spiky crystals (Figure 102), as in volcanic glass (Figures 81 and 82).

When we find glassy rocks of this kind along faults in very old rocks, we can infer that deep earthquakes affected those rocks in the past. However, because similar glassy rocks can be formed by melting caused by the impact of large meteorites (Chapter 11), geologists need to take care when working out the origin of these rocks.

Figure 98.

Fragmental rock ("fault breccia") formed by breaking of rock in a fault, followed by deposition of minerals between the fragments to make a solid rock. This rock makes a spectacular building stone, as in this shopping arcade in Boston, United States.

Figure 99.

Thin layers of calcium carbonate (calcite) repeatedly deposited from water-rich solutions in a cavity formed where a fault changed orientation, in a situation similar to that shown in Figure 97. Some wall-rock fragments (grey) are shown at the bottom-right of the sample. The black material is natural bitumen that has flowed into cracks. Santa Barbara, California, United States. The specimen is 20 centimetres across.

Sample courtesy of Rick Sibson.

Figure 100.

This dark grey, glassy rock formed in cracks as a result
of earthquake activity (fault movement) about 550
million years ago in the Musgrave Ranges, central
Australia. The knife is 9 centimetres long.

Figure 101.

Microscope photo of a melted rock from the same locality as the rock shown in Figure 100, showing radiating spiky outgrowths on colourless plagioclase and abundant feathery crystals of pyroxene (shades of grey). Normal colours; base of photo 0.7 millimetres.

Figure 102.

Microscope photo of a melted rock from the same
locality as the rock shown in Figure 100. The rock is
composed mainly of glass with abundant bubbles,
together with some unmelted mineral fragments
(colourless). Also shown are radiating spiky aggregates
that probably grew in the solid glass in a similar way to
the radiating aggregates that grow in volcanic glass
(Figures 81, 82). Normal colours; base of photo
4.4 millimetres.

All washed up

Rocks exposed to Earth's atmosphere

Rocks formed in Earth's crust and mantle and later exposed at the surface interact with the atmosphere because the minerals of the rocks are formed at higher temperatures and/or pressures and at drier conditions than those at Earth's surface. Chemical reactions between the minerals in the rock and the water and other chemical compounds in the atmosphere occur, resulting in the formation of new minerals that are more stable in the lower-temperature, wet conditions. We say that the rocks "weather."

These "weathering" reactions gradually change the old minerals to new minerals richer in oxygen (oxidation reactions) and water (hydration reactions). The main new minerals are hydrated iron oxides ("rust") and the **clay minerals** (a group of very fine-grained, water-rich, complex aluminosilicates with various other elements, especially potassium, magnesium, and iron). Some of the old minerals become dissolved in rainwater and may be precipitated elsewhere. These processes of reaction and solution, assisted by biological processes, cause the breakdown (erosion or weathering) of rocks to soil, which is so important to us and other animals for vegetation and hence our food. Soil consists essentially of clay minerals, together with sand or silt – small fragments of undissolved original minerals, especially **quartz,** which is one of the few common minerals formed in the crust that is stable at Earth's surface.

The loose material (sediment), formed by the breakdown of rocks, slides and is washed down slopes and eventually is deposited in low areas, such as river floodplains, coastal fringes, and the sea floor.

In mountainous areas where the rain runoff and hence the erosion rate are very high, rock fragments of various sizes may accumulate with the sand and soil in the lower regions, and may be transported in rivers towards the sea. Rubbing of the rock fragments together during this transport tends to wear away their sharp corners, eventually producing rounded boulders and pebbles. This process may be assisted by wave action in the surf, if the fragments reach the sea. Deposits of pebbles are common in rivers, especially near the base of high mountains, but also occur in some near-shore marine deposits.

If sediment is transported for relatively long distances in rivers and especially if it spends a long time in the surf zone along beaches, the fine-grained clay minerals are washed out from between the sand and pebbles and deposited further away – either towards the edges of river floodplains or further out to sea. Currents tend to sort the sediment into distinct sizes, such as pebbles (> 2 millimetres in diameter), sand ($2-\frac{1}{16}$ millimetres in diameter), silt ($\frac{1}{16}-\frac{1}{256}$ of a millimetre in diameter), and clay ($< \frac{1}{256}$ of a millimetre in diameter).

Figure 103.

Microscope photo of a sandstone, showing fragments of quartz and feldspar that were rounded by abrasion. The fragments were later cemented together by much smaller crystals of quartz (silicon dioxide) that were precipitated from water circulating between the sand grains. Interference colours; base of photo 1.3 millimetres.

Sedimentary rocks

Loose sediment is deposited in layers (beds), one on top of the other, in subsiding areas of Earth's crust. As the sediment becomes buried, it is compacted, and groundwater begins to circulate through the spaces between the particles. Chemical compounds dissolved in this water may be precipitated as new minerals ("cement") between the fragments, filling the spaces and forming a solid *sedimentary rock* by cementing the particles together (Figures 103, 104).

Sedimentary rocks composed mainly of pebbles, sand, and clay are called **conglomerates** (Figure 105), **sandstones** (Figures 103, 104), and **shales** (Figure 106), respectively.

Examination of the pebbles in conglomerates and the sand grains in sandstones may help us to work out what types of original rock were broken down to form the sediment. For

Figure 104.

Microscope photo of a sandstone, showing fragments of volcanic rock that have been rounded by abrasion in rivers and waves. The fragments show that volcanoes were eroded to produce the sediment. The fragments were later cemented together by very large crystals of calcite (calcium carbonate) that were precipitated from water circulating between the sand grains. Calcite is normally colourless in the microscope, but this rock has been stained pink with an organic dye to distinguish the calcite from other carbonate minerals that may be present. Normal colours (apart from the dye); base of photo 4.4 millimetres.

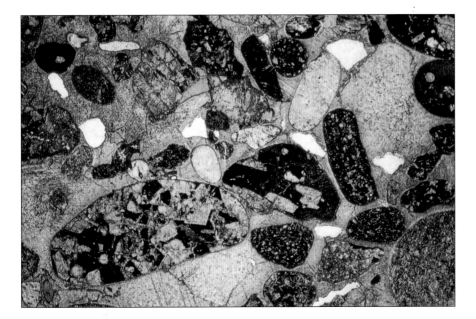

Figure 105.

Conglomerate (about 350 million years old) from Lake Keepit, northwestern New South Wales, Australia. The pebbles have been rounded by abrasion in a swiftly flowing river and/or by waves along a shoreline. They consist of several different types of rock, including granite. Because of the size and variety of the pebbles, conglomerates like this give a good idea of what types of rock were eroded to produce the sediment. The knife is 9 centimetres long.

example, the rounded sand grains in Figure 104 are mainly fragments of volcanic rocks, which tells us that volcanoes were eroded to produce the fragments. On the other hand, many of the pebbles in the conglomerate of Figure 105 are of granite, which tells us that rocks from deeper in the crust were uplifted and eroded to form the sediment.

Layered sequences of sedimentary rocks have been accumulated, buried, and later elevated by Earth movements throughout the known history of Earth's crust. Sedimentary beds are approximately horizontal when deposited, and may remain horizontal even when uplifted (Figures 107, 108). However, they commonly are tilted and folded by Earth movements when buried deep in the crust (Chapter 10), forming spectacular outcrop patterns when exposed millions of years later by erosion (Figure 14).

(text continues on page 124)

Figure 106.

Shale (about 350 million years old) from the Rocky Mountains, Canada. The thin layering (bedding) is typical of shales. Shales are the "ugly ducklings" of rocks, but when heated deep in Earth's crust, they change into beautiful "swans" like the rocks shown in Figures 127, 148–151, 155 and 156. The sample is 28 centimetres across.

Figure 107.

These horizontal beds of sandstone, shale, and limestone were
deposited on a coastal plain and beaches from about 220 to 570
million years ago, and uplifted to form a plateau (the Colorado
Plateau) about 65 million years ago. Erosion of the overlying, even
younger rocks (remnants of which are shown in Figure 108) has
produced a flat surface that has been deeply dissected by rivers,
producing canyons, the most famous being the Grand Canyon of the
Colorado River, Arizona, United States, shown here.

Photo by Scott Johnson.

Figure 108.

Horizontal beds of limestone, sandstone, and shale in Bryce Canyon National Park at the eastern edge of the Paunsaugunt Plateau in southern Utah, United States. Erosion of these rocks has produced spectacular turret-, pinnacle-, and fin-like outcrops.

Photo by Scott Johnson.

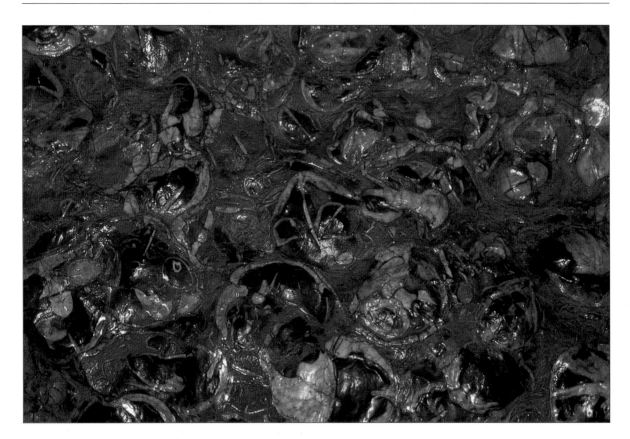

Figure 109.

Abundant fossil brachiopod shells in a limestone, about 400 million years old, from the Broken River area, northern Queensland, Australia. The sample, which is 82 centimetres across, has been lacquered for display purposes.

Courtesy of the Macquarie University Centre for
Ecostratigraphy and Palaeobiology.

Fossils

Rocks composed mainly of calcium carbonate (calcite) are called **limestones.** The calcium carbonate can be deposited directly from seawater to form rock layers, but usually is deposited in the shells of marine animals. As the animals die, their shells accumulate to form rocks consisting mainly of shell fossils (Figures 109, 110) or the shells may be broken into fragments (Figure 111). Later the shells and fragments are cemented together, generally with more calcite (Figure 111).

Some animals, such as corals, grow as colonies in reefs (for example, the Great Barrier Reef off northeastern Australia) and when the reef dies and is buried, fossil-bearing limestone is the result.

Vertebrate animals may also leave their impressions in soft material, such as clay, which hardens into shale (Figures 112, 113). Plants may also leave impressions of leaves and stems in shale (Figure 114).

Figure 110.

Well-preserved fossil mollusc shells, 370 million years old, in limestone from the Canning Basin, north Western Australia. The shells have been aligned by water currents. Also present are many small fragments of crinoids ("sea lilies"), which have skeletons made up of calcium carbonate plates. The specimen is 39 centimetres across.

Display sample courtesy of the Macquarie University Centre for Ecostratigraphy and Palaeobiology.

Figure 111.

Microscope photo of limestone (about 350 million years old), Lake Keepit area, northwestern New South Wales, Australia. The limestone is composed largely of fossil shell and crinoid ("sea-lily") fragments that probably accumulated in a lagoon near a coral reef being abraded by wave action. The fragments have been cemented together by calcite. The dark rims on the fragments were formed by very fine-grained calcium carbonate "mud" sticking to them as they were gently washed about in the lagoon. Normal colours; base of photo 1.8 centimetres.

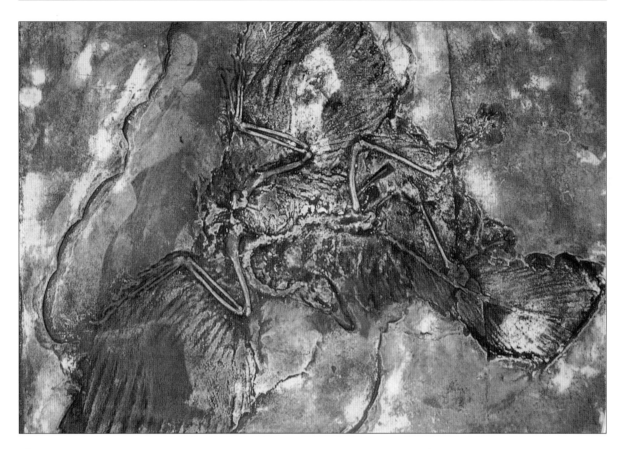

Figure 112.

Plaster cast of a beautifully preserved specimen of *Archaeopteryx,* made from an impression preserved in shale about 200 million years ago. The large, powerful legs are like those of some of the smaller two-legged dinosaurs, but the feathers show that it was either the earliest known bird or a small, transitional, bird-like dinosaur.

Display sample courtesy of the Macquarie University Centre for Ecostratigraphy and Palaeobiology.

Figure 113.

Excellent impression of a fish in shale, about 50 million years old, from Wyoming, United States. The fish is 11 centimetres long.

Display sample courtesy of the Macquarie University Centre for Ecostratigraphy and Palaeobiology.

Figure 114.

Plant leaf impressions in shale, 270 million years old, from Dunedoo, New South Wales, Australia. The plant is *Glossopteris* and the slab is 28 centimetres across. The leaves were pressed into soft clay, which eventually hardened.

Display sample courtesy of the Macquarie University
Centre for Ecostratigraphy and Palaeobiology.

8

Turning up the heat

How do we know about deep Earth processes?

We observe that earthquakes occur (indicating faulting at depth) and that lava is erupted in volcanoes (indicating melting of rock at depth). We also know that some parts of Earth's surface are rising rapidly – about 5 millimetres a year, which is very fast, geologically speaking. The Himalayas and the Southern Alps of New Zealand are good examples.

The resulting mountain ranges are rapidly eroded by the weather. This erosion removes overlying rocks and eventually exposes the uplifted deeper rocks. This is how we can observe minerals and structures that were formed and modified at great depth in Earth's crust many millions of years ago.

It's hard to imagine how huge volumes of Earth's crust can be elevated from depths of 40 kilometres or more to become exposed at the surface. However, when massive blocks of lithosphere crash into each other as plates move (Figure 12), the crust becomes folded and fractured (Figures 19, 20, 21). Eventually, enormous slices of it are pushed and squeezed outwards and upwards (Figure 13). The Alps were built up in this way (Figure 14).

Why are minerals formed at depth preserved at Earth's surface?

As discussed in Chapter 7, when rocks formed deep in the crust at high temperatures and pressures are exposed at the surface, they react with the water and gases in the atmosphere. Because the atmosphere is much cooler than the crust or mantle where the minerals were formed, the minerals are unstable at Earth's surface. Why doesn't this "weathering" destroy all evidence of the minerals formed at depth in the crust?

The reason is that weathering is a very *slow* process, because chemical reactions need heat to proceed rapidly and the cool atmosphere slows them down. The fortunate result is that the original minerals and structures are well preserved underneath a generally thin rind of weathered rock. The preservation in exposed rocks of minerals and structures formed at great depth long ago provides us with a wonderful opportunity to interpret the history of ancient Earth rocks.

Experiments

Another way of obtaining information about processes that occur deep in Earth's crust and mantle is by carrying out laboratory experiments at high temperatures and pressures on mixtures of chemical compounds identical to those of actual minerals. By subjecting these compounds to various temperatures and pressures, experimenters learn what minerals are stable at different pressure-temperature conditions. Therefore, when we find the same minerals in natural rocks, we have a good idea of the pressure-temperature conditions to which they were subjected in Earth's crust or mantle.

The experiments show that some parts of Earth's ancient crust have been subjected to temperatures of up to 1,100° C and pressures of up to 30 kilobars (3,000 megapascals) for millions of years. Temperatures of 400–800° C and pressures of 5–8 kilobars – equivalent to a depths of 17–26 kilometres – are commonly indicated by the minerals in large volumes of the deeper parts of Earth's crust.

Changes in solid rocks deep in Earth's crust

Rocks formed at Earth's surface (such as sedimentary and volcanic rocks), as well as rocks formed below the surface (such as intrusive igneous rocks), are all subjected to Earth movements, and eventually may become buried to great depths in the crust.

As a result, minerals formed much higher in the crust or at Earth's surface become unstable, and thus change to new minerals that are more stable at the higher temperatures and pressures. This process of mineral change is called *metamorphism* and the resulting rocks are called *metamorphic rocks*. It is important to appreciate that the mineral changes occur in *solid rock*, apart from small amounts of watery fluids in small cracks.

How do we know that rocks remain solid during metamorphism? The answer is that though metamorphism tends to obliterate original igneous or sedimentary structures, many rocks in which the original minerals have been completely changed to new minerals retain original structures. This applies especially to rocks that have not been deformed very much during the metamorphism.

For example, metamorphosed shales and sandstones may preserve sedimentary bedding (Figures 115, 116) and conglomerates may retain the shapes of pebbles – typically stretched and/or folded by the deformation that accompanies the metamorphism (Figures 117, 146, 147). Similarly, metamorphosed igneous rocks may preserve the shapes of residual gas bubbles, now filled with new minerals (Figure 62), or the shapes of igneous crystals completely or partly replaced by new metamorphic minerals (Figure 118).

These residual structures (now outlined by new metamorphic minerals) are helpful in working out what the rock was before metamorphism, which can be difficult to do when examining strongly metamorphosed rocks. This information is essential for deciphering the history of the deeper parts of Earth's crust. It is also useful when searching for potential ore-bodies in metamorphic rocks, because different associations of ore minerals occur in intrusive igneous, volcanic, and sedimentary rocks.

An example of the preservation of original structures is shown by a limestone containing fossil shells (similar to that shown in Figure 109) that was heated by an intrusion of granite magma 5–10 kilometres below the surface. Where the rock was metamorphosed at relatively low temperature (well away from the intrusion), the calcium carbonate in the shells simply grew to form much larger, polygonal grains of calcite (Figure 119). However, where the rock was metamorphosed at higher temperature (close to the intrusion), the shells were converted to calcium silicate (wollastonite), as shown in Figure 120. In both Figures 119 and 120, the curved and tapered shapes of the cross sections of the original shells can be observed, despite the growth of new mineral grains.

It's worth noting that the growth of the new calcite did not involve any chemical change, as the shells originally consisted of calcium carbonate. The only changes involved grain size and grain shape. However, the growth of wollastonite did involve a chemical change. The calcite of the shells reacted with quartz (silicon dioxide) grains between the shells to form

(text continues on page 140)

Figure 115.
Sedimentary bedding preserved in metamorphosed
shales and sandstones, about 450 million years old,
Slacks Creek, Cooma, southeastern New South Wales,
Australia. The knife is 9 centimetres long.

Figure 116.

Sedimentary bedding preserved in metamorphosed shales and limestones, King Island, Tasmania, Australia. Exchange of chemical elements between the thin shale and limestone beds during the heating has produced layers of different metamorphic minerals with strikingly different colours, accentuating the bedding. The minerals are pyroxene (green), garnet (brown), and biotite mica (dark). The knife is 9 centimetres long.

Figure 117.

Deformed, metamorphosed conglomerate from the
Alps. The pebbles have been stretched by flow during
heating and deformation of the rock, which would
have resembled the conglomerate shown in Figure 105
before the heating and deformation.

Photo by Scott Paterson.

Figure 118.

Microscope photo of a metamorphosed gabbro,
Adirondack Mountains, New York, United States.
Many of the elongate shapes of the original igneous
plagioclase crystals (shades of grey) are still visible,
though the original large pyroxene crystals have been
completely replaced by aggregates of polygonal grains
of metamorphic amphibole. Interference colours; base
of photo 1.8 centimetres.

Figure 119.

Microscope photo of a metamorphosed limestone,
about 380 million years old, showing the curved and
tapered shapes of cross sections of fossil shells. The
shells have been recrystallized to aggregates of
polygonal calcite grains that are much larger than the
original calcite in the shells. Cox's River, near Hartley,
west of Sydney, New South Wales, Australia.
Interference colours; base of photo 1.8 centimetres.

Figure 120.
Microscope photo of a metamorphosed limestone, about 380 million years old, showing the curved shapes of cross sections of fossil shells. The shells have been replaced by aggregates of fibrous grains of wollastonite (calcium silicate), because of a chemical reaction between the shells and grains of quartz in the fine-grained material between the shells. Cox's River, near Hartley, west of Sydney, New South Wales, Australia. Interference colours; base of photo 1.8 centimetres.

calcium silicate, liberating carbon dioxide as a gas, which escaped from the rock. By simple chemical reactions like this, as well as more complicated reactions, new minerals are produced during metamorphism.

New minerals

Metamorphic minerals reflect the *chemical composition of the rock* in which they are growing. For example, shales have abundant clay minerals (Chapter 7), which are rich in aluminium. Therefore, metamorphism of shales produces new minerals rich in aluminium, such as mica and garnet. Similarly, limestones contain fossil shells and corals that are rich in calcium. Therefore, metamorphosed limestones contain minerals rich in calcium, such as wollastonite.

Metamorphic minerals also reflect the *temperature* and *pressure* to which the rocks were subjected in Earth's crust. For example, we have just seen that calcite and quartz occur together in areas of weak metamorphism, whereas wollastonite grows instead if the rocks are hotter.

New structures

Minerals in sedimentary rocks tend to have fragmental shapes, whereas minerals in igneous rocks tend to have either crystal faces or shapes determined by spaces available for them when they crystallize during cooling of the magma.

However, minerals in metamorphic rocks all grow at the same time, as products of chemical reactions that occur in response to the heating. Because they grow simultaneously in solid rock, the new minerals adopt shapes that reflect this process. They grow until they impinge on one another and then change their shapes, not only to fill the volume of the rock, but to *minimize the energy of the boundaries between them.* This energy is caused by the fact that the atoms at the boundaries where different mineral grains meet are not able to fully join one grain or the other, and so are in less regular ("higher-energy") positions than they would be if they were regularly organized inside a grain or crystal.

Many metamorphic minerals minimize this grain-boundary energy by reducing their grain boundary area. The best way for them to achieve this reduction in area would be to form hexagons in two dimensions, as bees do in honeycomb. However, this requires evenly distributed growth centres, which is most unlikely in rocks, in which the old minerals (in which the new ones grow) are generally distributed unevenly. The minerals compromise by forming polygons with three to seven sides (in two dimensions), the most common shapes being pentagons (Figures 121–123). Three grains meet at a point in two dimensions (four

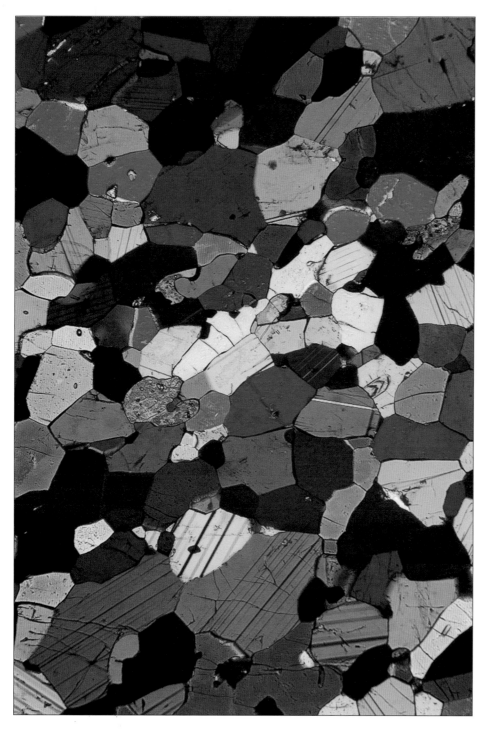

Figure 121.

Microscope photo of polygonal grains of plagioclase (shades of grey) and epidote (bright colours) in a metamorphic rock from Mount Painter, South Australia. Interference colours; base of photo 4 millimetres.

Photo by Graham Teale.

Figure 122.
Microscope photo of polygonal grains of plagioclase (shades of grey) in a
metamorphic rock (about 1,700 million years old) from Broken Hill, western New
South Wales, Australia. The different colour intensities separated by straight lines
inside the plagioclase indicate that parts of the grains grew in different orientations.
They are called "twins." Interference colours; base of photo 4.4 millimetres.

grains at a point in three dimensions). Ideally, the boundaries of the grains should meet at
angles of 120° in two dimensions (109° 28' in three dimensions), to minimize the grain-
boundary area, but the actual angles may be a little different from these theoretical values,
owing to variations in the arrangements of the atoms in different directions in the minerals
involved.

Polygonal grain shapes are common not only in minerals formed by metamorphism in
Earth's crust, for example plagioclase (Figures 121, 122), but also in other "hot" materials,

such as olivine in peridotite in Earth's mantle (Figures 43, 123) and glacier ice on Earth's surface (Figure 38). Polygonal shapes are also common in many ore deposits (sulphide and oxide minerals), and are typical of organic cells, annealed metals, and many synthetic ceramic materials.

However, not all minerals are able to form polygonal aggregates, because the bonding between their atoms is so strong in some directions and so weak in others that their boundaries cannot easily adjust to boundaries of other minerals. Instead, these minerals reduce the energy of their boundaries by forming crystal faces, which tend to have lower energy than random boundaries. The most common example is mica (Figures 124, 125). The contrast between the polygonal grains of plagioclase and the crystal shapes of mica is well illustrated in Figure 124. Another mineral that typically forms elongate crystals in metamorphic rocks is wollastonite (Figures 120, 126).

Figure 123.
Microscope photo of polygonal grains of olivine in a rock from Earth's mantle, brought up as a fragment in a basalt lava flow on the island of Hawaii. Interference colours; base of photo 4.4 millimetres.

Figure 124.
Microscope photo of a metamorphic rock, 1,700 million years old, from Broken Hill,
western New South Wales, Australia, showing polygonal grains of plagioclase (grey)
interspersed with crystals of mica (bright colours). Interference colours; base of
photo 4.4 millimetres.

Many other minerals, such as garnet (Figure 126) and chloritoid (Figure 127), develop
crystal faces in metamorphic rocks under certain conditions. The reason for this is not yet
clear, but may be due to the presence of very thin films of fluid along the crystal boundaries,
which permits the development of crystal faces, as liquid magma does in igneous rocks.
Some crystals grow abnormally large in metamorphic rocks (Figure 128).

(text continues on page 148)

Figure 125.

Crystals of mica (muscovite, with bright colours) in a
metamorphic rock, about 1,700 million years old,
from Broken Hill, western New South Wales,
Australia. Note the numerous straight crystal faces.
Interference colours; base of photo 1.75 millimetres.

Figure 126.
Metamorphosed limestone (about 380 million years old) from Duckmaloi, near Oberon, New South Wales, Australia, showing radiating, lustrous white crystals of wollastonite and large black crystals of calcium-iron garnet. The base of the sample is 30 centimetres across.

Figure 127.
Microscope photo of a metamorphosed shale (about 1,200 million years old) containing large, pale bluish-green crystals of chloritoid (iron aluminosilicate), Ontario, southeastern Canada. Note the straight crystal faces of the chloritoid. Originally this rock would have resembled the shale of Figure 106. Normal colours; base of photo 4.4 millimetres.

Figure 128.
Very large crystals of red-brown garnet in a metamorphic rock (about 1,200 million
years old) in the Adirondack Mountains, New York, United States. The garnet (itself
formed by a metamorphic reaction) has undergone a later chemical reaction with
surrounding minerals, producing dark rims of amphibole on the garnet crystals.
Therefore, this rock preserves a history of different metamorphic events.

Sequences of change

Another way of obtaining evidence on deep Earth processes is to observe a progressive
change from unheated to heated rock (as now exposed at the surface).

Temperatures and pressures obviously increase with depth in Earth's crust. However,
they can be very variable from place to place, even at the one crustal level. This is mainly
because the amount of heat that enters the crust from the mantle is variable. Some heat is

transferred to the crust by direct upsurge of the mantle itself (for example, as portions of the lower crust collapse into the mantle), and some heat is transferred by rising basalt magma. This very hot basalt magma can melt rocks of the deeper crust, and the resulting granite magma (Chapter 5) carries heat even higher.

Therefore, though the deeper parts of the crust are hot and are continually being strongly metamorphosed, the heating of the middle crust can be quite variable. It is heated and squeezed on a regional scale (tens of kilometres), but some parts of it are heated more than others. Thus, we can observe the resulting transition from less to more strongly altered rocks after the region has been uplifted and exposed at Earth's surface, many millions of years later.

For example, the change of gabbro, like that shown in Figure 55, to a granular metamorphic rock with the same minerals is shown in Figures 129 and 130. The elongate igneous crystal shapes are gradually replaced by polygonal metamorphic grain shapes.

Figure 129.
Microscope photo of a metamorphosed gabbro, about 1,800 million years old, from the Anmatjira Range, central Australia. Though most of the grains are polygonal, owing to the metamorphism, the rock retains some of the elongate shapes of the original igneous plagioclase. Interference colours; base of photo 4.4 millimetres.

Figure 130.
Microscope photo of a completely metamorphosed gabbro, in which
all grains are polygonal. The minerals are plagioclase (shades of grey)
and pyroxene (shades of brown, yellow, blue and green). Interference
colours; base of photo 4.4 millimetres.

Melting

If the temperature of metamorphism is high enough (greater than 650° C), many rocks
begin to melt. The melted portion migrates through the rock and collects to form light-
coloured patches, veins, and layers (Figure 131). Partly melted rocks tend to flow extensively
in response to Earth forces, forming intricate folded patterns. This is partly because of the
weakness of the melt itself and partly because even the unmelted minerals are so hot that
they flow easily.

As the temperature rises and melting becomes more extensive, the melt may coalesce even more, and may move small distances to form small, local granite intrusions (Figure 132). Melting at still higher temperatures (greater than 850° C) produces even larger amounts of melted material, which coalesces and rises as larger bodies of magma that form very large intrusions of granite in the middle and outer crust.

Figure 131.

Strongly heated, strongly folded, partly melted metamorphic rock in the Darling Range, Broken Hill, western New South Wales, Australia. The light-coloured material (quartz and feldspar) is former melt that has segregated into veins and solidified. The darker material is unmelted rock. The original rock was a shale (like that shown in Figure 106), apart from a bed of sandstone, which is preserved as a strongly folded, light-grey layer in the centre of the outcrop. The knife is 9 centimetres long.

Figure 132.
Local segregation of melted rock (light colour), Helsinki, Finland. The molten rock
penetrated along and across the rock's foliation. Both the light and darker
(unmelted) layers were folded by Earth movements while they were still hot, but after
the melted rock had crystallized. Backpack for scale.

As mentioned in Chapter 5, this regional-scale melting is generally triggered by hot
basalt magma from Earth's mantle, which not only melts crustal rocks to form granite
magma, but mixes and mingles with the granite magma itself (Figure 69).

How can new minerals grow in solid rocks?

Many of us are familiar with crystals growing in liquids, but the idea of crystals growing in
other solid, crystalline minerals may be difficult to understand. Two main processes are

probably involved, namely: (1) movement ("diffusion") of atoms inside crystals and along the boundaries between them, and (2) dissolving of old minerals and precipitation of new minerals, a little bit at a time, in the very small amount of fluid that is generally present (Chapter 9). For example, many of the old minerals contain water, which is released when they are heated to form new minerals that contain less or no water. Metamorphism probably involves both diffusion and solution-precipitation under most conditions in Earth's crust.

In hot water

Water, water everywhere . . .

If we drill down a little way into solid rock, water may come up the hole. This tells us that rocks beneath the surface can contain water in cracks and cavities. A well-known example is artesian water, which commonly occurs in the spaces between fragments in sandstones that have not been thoroughly cemented together (Chapter 7). The caves and underground pools in very soluble rocks, such as limestone, are also evidence of the activity of subterranean water.

Moreover, water has been found in faults intersected by even the deepest drill holes, at 9–11 kilometres below the surface, in Russia and Germany. Therefore, we can infer that water may be present almost everywhere in the rocks of Earth's crust, especially in fault zones.

Water in magma

Many of the minerals that crystallize from magma as it cools contain no water (for example, olivine, pyroxene, feldspars, and quartz). Therefore, water tends to concentrate in the liquid left between the growing crystals. If the magma is extruded onto Earth's surface as lava, this water boils off, locally leaving bubble holes (Figures 61, 62, 84, 85).

155

However, if the magma is trapped in Earth's crust, the water is released only when the rock contracts as it solidifies to form cooling cracks, into which the water can escape. The water may then react with the higher-temperature, dry igneous minerals, converting them to lower-temperature, water-bearing minerals (Figure 133) and also depositing such minerals in any cavities that may be present in the igneous rock (Figures 61, 62, 134).

Water in rocks at depth in Earth's crust is naturally hotter than surface water. However,

Figure 133.

Microscope photo of a gabbro from Cyprus that has been altered by hot water. The original pyroxene (as shown in Figure 55) has been changed to fine-grained fibrous, water-rich minerals (mainly amphibole), but the plagioclase has escaped alteration. Interference colours; base of photo 2.5 centimetres.

Figure 134.
Microscope photo showing radiating, fibrous, water-bearing minerals
(prehnite and pumpellyite) that grew into a cavity in basalt attacked
by hot, water-rich solutions. Later the rest of the cavity was filled with
calcite. Interference colours; base of photo 4.4 millimetres.

water in rocks heated by intrusions of magma is especially hot, and so is a potential source of
heat energy, which has been tapped in several places (for example, in the North Island of New
Zealand). This kind of heating is especially common in volcanic areas, such as Yellowstone in
the United States (Figure 135) and Rotorua in New Zealand (Figure 136). In these areas, magma
trapped below the surface heats the water, which reaches the surface as geysers and hot springs.
Quartz and opal may be deposited in layers ("terraces") on the surface (Figure 135), indicating
that the hot water dissolves silicon dioxide from the rocks through which it passes.

The hot water may also dissolve metals and sulphur from the rocks through which it cir-
culates. The metals may be deposited later as metal sulphides or sulphur in rock cavities

Figure 135.

Terraces of "geyserite" (a mixture of opal and quartz) deposited from hot spring
waters at Yellowstone National Park, Wyoming, United States.

Photo by Scott Johnson.

(Figures 137, 138, 139) or on Earth's surface (Figure 136). Evidence of the action of hot water
in old volcanic rocks can be an indication to geologists that a potentially valuable metal
deposit is nearby.

Water, faults, and ore deposits

As noted in Chapter 6, an open space may be produced as a fault moves. Water travelling along
the fault may deposit minerals in the open space (Figures 96, 98, 99). In addition, water in
cracks in the adjacent rocks is sucked into the opening because of the lower pressure (Figure
97). This water may deposit minerals from chemical compounds carried in solution. It may

Figure 136.
Hot spring at Rotorua, New Zealand, showing a cloud of condensing steam and
deposits rich in "geyserite" (a mixture of opal and quartz), stained with yellow sulphur.

also mix with water travelling along the fault, and the resulting chemical reactions may cause
metal sulphide minerals to be deposited in the cavity (Figures 137, 138, 139), including the spaces
between any rock fragments present. These minerals cement the fragments together, forming a
fault breccia (Figures 98, 99). This is a major potential source of metallic ore minerals.

Water and ore deposits at mid-ocean ridges

Circulating seawater becomes heated by volcanic activity and dissolves metals from hot
basalts accumulating at mid-ocean ridges (Figure 11) as it percolates through them along
cracks. The water then deposits metal sulphides when it is released back into the cold ocean.

(text continues on page 162)

Figure 137.

Crystals of pyrite (iron sulphide) with a metallic lustre, deposited in a
cavity with crystals of quartz, Cananeo, Mexico. Specimen 18
centimetres long.

Macquarie University display sample.

Figure 138.

Metallic sulphides deposited from solution in hot water between rock fragments in a breccia, Mount Gunson, South Australia. The sulphide minerals are chalcopyrite (copper sulphide, brassy in appearance) and sphalerite (zinc sulphide, shiny dark grey). The specimen is 18 centimetres long.

Macquarie University teaching sample.

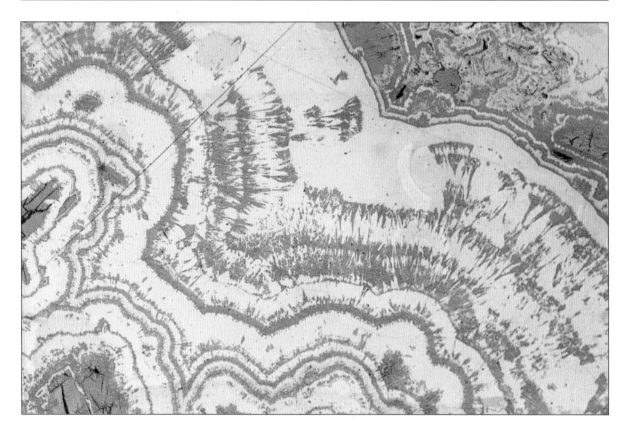

Figure 139.
Microscope photo of a polished surface of a rock from the same locality as the rock in
Figure 138, showing sulphide minerals deposited in layers in a cavity. The layers are
built up, one on the other, by repeated deposition from hot water. Some of the
minerals form radiating fibrous crystals that grew from the edges of the cavity towards
the centre. The yellow mineral is chalcopyrite (copper iron sulphide), the purplish
grey mineral is sphalerite (zinc sulphide), and the blue mineral is covellite (copper
sulphide). Because most sulphide minerals are opaque, even in very thin slices, they
are examined by reflecting light from polished surfaces. Base of photo 2.7 millimetres.
Specimen courtesy of John Lusk.

The sulphide minerals are deposited as very small, black particles resembling black smoke
(Figure 140) and build up local accumulations on the sea floor, forming "chimneys" (known
as "black smokers") around the vents through which they are released (Figures 140, 141).
Because some very large ore deposits in older rocks (now exposed at Earth's surface) may
have been formed in this way, geologists search for rocks that may indicate a former oceanic
setting of this type.

Figure 140.

Pencil-like "black smoker" tubes on the side of a large chimney structure covered with baseball-sized snails, in the Pacmanus hydrothermal field, about 1,700 metres below sea level, Manus Basin, Bismarck Sea. The tubes are made largely of copper sulphide minerals. The black "smoke" emanating from the tubes consists of metallic sulphide particles deposited from solution in seawater at a measured temperature of 280° C.

The photo is a videomosaic prepared by the Commonwealth Scientific and Industrial Research Organization from a recording by Ray Binns taken from the "Shinkai-6500" submersible belonging to the Japan Marine Science and Technology Agency.

Figure 141.

Large "smoking" chimney complex, about 3 metres high, made of copper-rich and zinc-rich sulphides and carrying high levels of gold and silver, in the Pacmanus hydrothermal field, about 1,700 metres below sea level, Manus Basin, Bismarck Sea.

The photo is a videomosaic prepared by the Commonwealth Scientific and Industrial Research Organization from a recording by Ray Binns taken from the "Shinkai-6500" submersible belonging to the Japan Marine Science and Technology Agency.

A tight squeeze

Changing the shapes of solid rocks

Earth forces have been mentioned previously, for example, in connection with plate motion, continent collision, and the elevation of mountain ranges. The huge forces that rocks are subjected to in zones where lithospheric plates slowly collide or rub past each other cause the rocks to change not only their positions with respect to each other, but also their shapes.

As rocks heat up while being squeezed, they slowly change their shapes by flow (Chapter 3), which is accompanied by growth of new grains of the same minerals (recrystallization). At the same time, new minerals grow in response to increasing temperatures (Chapter 8). The new minerals tend to grow with their long dimensions aligned in the plane in which the rocks are being flattened. The result is a strong alignment of minerals in a "foliation" (from "folium," the Latin word for leaf), as shown in Figures 142–145.

Squeezing and shearing of hot rocks that are layered (for example, sedimentary rocks with bedding or metamorphic rocks with a foliation) produces buckles or *folds,* in much the same way as a sheaf of papers crumples if you squash it parallel to the sheets (Figures 144–148). A new foliation commonly develops in the folds at a high angle to the original layering (Figures 144, 145). As folds tighten in response to increasing intensity of Earth forces, the foliation intensifies (Figure 144).

(text continues on page 170)

Figure 142.

Microscope photo of well-aligned mica (mainly brightly-coloured muscovite, with smaller amounts of brown biotite) forming a foliation in a metamorphic rock from Nepal. The alignment is due to the deformation that accompanied heating. Interference colours; base of photo 4.4 millimetres.

Thin section courtesy of Scott Johnson.

Figure 143.
Folding of dark layers (rich in mica) and light layers
(rich in quartz and plagioclase) in a metamorphic
rock from Otago, New Zealand. The layering is not
sedimentary bedding, but is a foliation resulting from
the segregation of mineral components during
heating and deformation, which occurred about 140
million years ago.

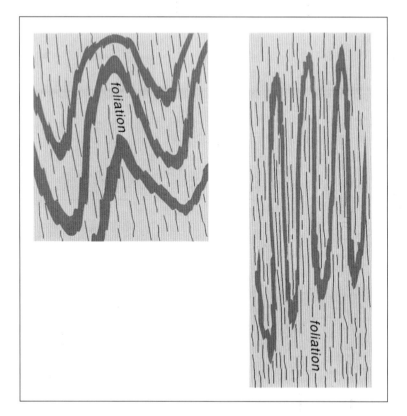

Figure 144.

Sketches showing the development of a foliation
(caused by alignment of new minerals) during folding
of sedimentary beds (pink layers). As the folds tighten
in response to increasing compression by Earth forces,
the foliation intensifies and becomes parallel to the
beds, which are squeezed thinner.

Figure 145.
Sedimentary rocks that were so strongly folded and
heated (about 1,600 million years ago) that a new
foliation, very oblique to the original layering, was
produced; Bonya Plain area, central Australia. The
foliation is caused by the aligned growth of new
minerals, in response to the Earth forces that caused
the folding, which occurred about 15 kilometres
beneath the surface at a temperature of around 600° C.
The pen (which is parallel to the new foliation) is 15
centimetres long.

Figure 146.

Outcrop showing evidence of two episodes of strong folding of a conglomerate (about 1.9 billion years old), on the west side of Amisk Lake in Eastern Saskatchewan, Canada. Several pebbles of different rocks have been stretched in one deformation event and then folded in a second event. In the centre of the photo is a large pebble of pink granite that was stretched and dismembered into lenticular pods during the first deformation episode. Later the stretched pebble was squashed parallel to its length and folded in the second deformation episode. Pen for scale.

Photo by Jim Ryan.

Many metamorphic rocks show evidence of multiple episodes of deformation (Figures 24, 146, 147, 148), reflecting a complicated history of Earth movements. Moreover, large crystals of metamorphic minerals may overgrow folded patterns and preserve them inside the crystals, even where later heating and deformation have removed all evidence of folds from the surrounding rock (Figures 149, 150, 151). Preserved patterns of this kind can help geologists work out quite complicated histories of heating and folding events that have occurred at different times in Earth's crust.

(text continues on page 175)

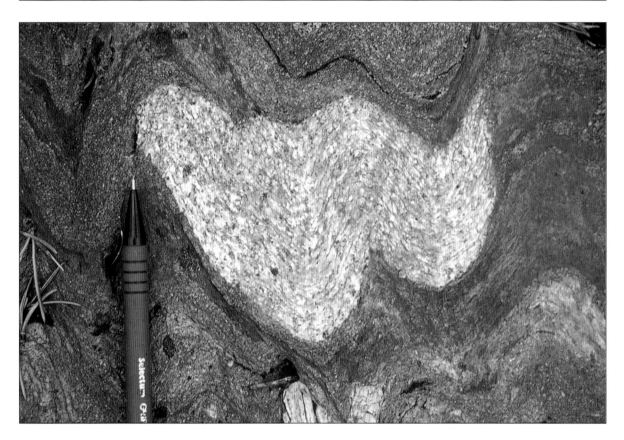

Figure 147.

Deformed pebble of granite at the same locality as shown in Figure 146. The pebble was stretched into an elongate lens shape during the first deformation episode, and folded in the second deformation episode. Compare this deformed granite with an undeformed granite shown in Figure 3. Pen for scale.

Photo by Jim Ryan.

Figure 148.

Microscope photo showing the result of two episodes of strong folding in a metamorphosed shale, Picuris Range, New Mexico, United States. The first episode produced a foliation that was re-folded ("crenulated") in the second episode, forming a new foliation marked by layers of strongly aligned crystals of mica (bright colours) very oblique to the first foliation. This rock originally resembled the shale in Figure 106, which shows how metamorphism can change rocks completely. Interference colours; base of photo 12.0 millimetres.

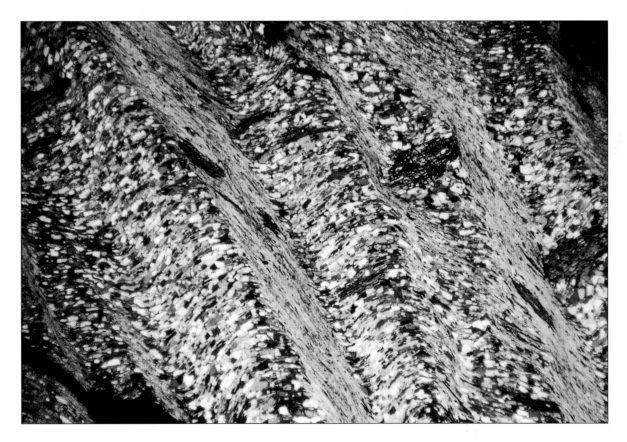

Figure 149.

Microscope photo showing a large crystal of staurolite (pale yellow) with folded trails of graphite (black), which were preserved by the crystal as it grew. Since then, new minerals and structures have developed in the surrounding rock, removing the evidence of the earlier folding. Preservation of evidence of earlier events in large crystals helps us to decipher the rock's history. Cascade Range, near Leavenworth, Washington, United States. Normal colours; base of photo 4.4 millimetres.

Figure 150.

Microscope photo showing spiral trails of small quartz grains (clear) enclosed in a large, irregularly shaped grain of garnet in a schist (metamorphosed shale) from the Himalayas in Nepal. The spiral resembles a spiral galaxy of stars, and generally is interpreted as being caused by rotation of the garnet crystal as it grew during deformation. However, other interpretations have also been suggested. Normal colours; base of photo 4.4 centimetres.

Thin section provided by Scott Johnson.

Figure 151.

Same view as in Figure 150, but with interference colours. This view shows the spiral clearly. Base of photo 4.4 centimetres.

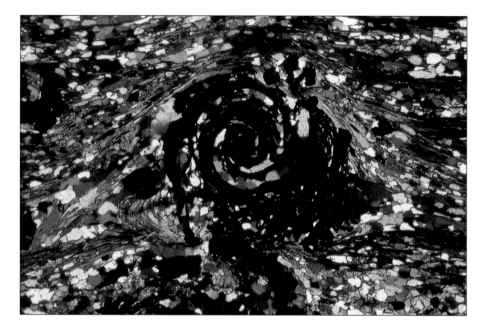

Progressive deformation and heating of shale in Earth's crust

When a shale is heated and squeezed in Earth's crust, its clay minerals become converted to new minerals, especially mica, which tend to grow in a preferred direction that reflects local Earth forces and so forms a "foliated" rock.

When shale is heated at low temperatures (300–400° C), mineral growth is slow. The typical foliated rock produced is a **slate,** in which the very small, flat mica crystals grow sub-parallel to each other, forming a "slaty cleavage" (Figures 152, 153). This term implies that the

Figure 152.
Folded shales, with thin beds of sandstone (about 500 million years old), at Bermagui on the south coast of New South Wales, Australia, showing a strongly developed, almost vertical foliation ("slaty cleavage") very oblique to the approximately horizontal sedimentary bedding.

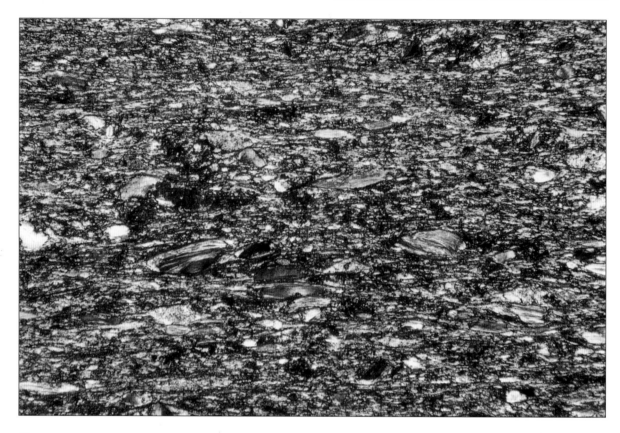

Figure 153.

Microscope photo of slate (shale metamorphosed at relatively low temperature, around 300–400° C). The mineral grains – mainly mica (bright colours) and quartz (white and grey) – are small and the mica crystals are well aligned. Interference colours; base of photo 1.75 millimetres.

slate can be split (cleaved) easily because of the strong alignment of the easily cleaved mica flakes. The cleavage makes slates useful for roofing and paving.

At moderate to high temperatures (400–700° C), mineral growth is faster, and thus the grains are larger. Moreover, the rocks (**schists**) do not split as cleanly as slates because the foliation is less regular. The reasons for this irregularity are the larger size of the crystals aligned in the foliation and the common presence of larger crystals of some metamorphic minerals, around which the foliation is deflected (Figures 150, 151, 154). Many schists have been deformed and folded several times, resulting in spectacular crenulations of the mica foliation (Figure 148).

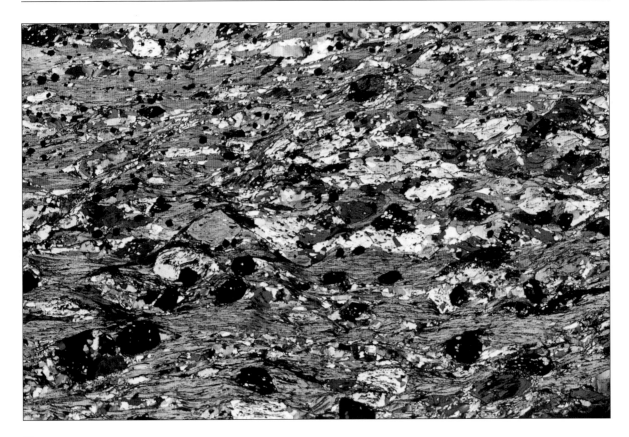

Figure 154.

Microscope photo of a metamorphosed sedimentary rock (schist) from Japan, containing abundant mica (muscovite) with a strong tendency to be aligned. The alignment is due to the deformation that accompanied the heating. The mineral grains (mainly mica and quartz) are larger than in slates (Figure 153), but not as well aligned. This rock started out resembling the shale in Figure 106, which illustrates the extent to which metamorphism can change rocks. Interference colours; base of photo 1.8 centimetres.

At higher temperatures of metamorphism (600–850° C), coarse-grained **gneisses** are formed. These rocks generally have very irregular, commonly discontinuous foliations, owing to the very large sizes of the metamorphic minerals (Figure 155).

A gneiss with minerals like sillimanite and feldspar (Figure 156) bears no resemblance to the shale (Figure 106) that it was originally. This is a striking example of how the growth of new minerals and the development of new structures during metamorphism can completely transform the appearance of the original rock.

Figure 155.
Microscope photo of a gneiss, showing a layer rich in sillimanite (needle-like
crystals) anastomosing around large grains of feldspar (grey) with quartz (grey) and
biotite mica (brown). The large feldspar grains, as well as the relatively large quartz
and biotite grains, prevent the formation of a close, strong foliation like those in
schists and slates (Figures 153, 154).

Zones of intense flow in Earth's crust

The flow of rocks in Earth's crust is very heterogeneous. Strongly deformed zones alternate
with weakly deformed zones (Figures 157, 158). Original structures tend to be better preserved
in the less deformed rocks, whereas they are generally obliterated in strongly deformed rocks.

The most strongly deformed rocks occur in relatively narrow zones called **mylonite**
zones. The word "mylonite" implies a "milled rock" (as though the deformation produced an
aggregate of small fragments), and some are just that, especially in cold rocks deformed high

Figure 156.

Microscope photo of gneiss from Broken Hill, western New South Wales, Australia.
Originally this was a shale (like that shown in Figure 106), which was metamorphosed
at high temperatures (about 700–850° C), about 1,600 million years ago. The rock is
composed of large grains of quartz and feldspar (shades of grey), crystals of biotite mica
(brown), and crystals of sillimanite (bright colours), which is an aluminosilicate
mineral formed at high temperatures. Interference colours; base of photo 4.4

in the crust. However, mylonites produced by deformation of hotter rocks deeper in the
crust generally flow, instead of fracturing. For example, Figure 159 shows deformation so
intense that formerly continuous layers have been squeezed into thin discontinuous lenses.
Nevertheless, evidence of flow is preserved in the form of abundant small folds. A rock even
more intensely deformed by flow is shown in Figure 160.

   A progressive change from undeformed to intensely deformed rock can be seen at the
edges of mylonite zones (Figure 157). This can be appreciated by comparing undeformed

(text continues on page 184)

Figure 157.

Zone of strong deformation (mylonite) marked by intense foliation in
a deformed granite (about 2,600 million years old) at Cape Carnot,
Eyre Peninsula, South Australia. Local mylonite zones like this are
common in strongly deformed metamorphic rocks. Pen (15
centimetres long) for scale.

Figure 158.

Intensely deformed granite, about 1,800 million years old, from Cape Donington,
Eyre Peninsula, South Australia, showing abundant, deformed residuals of feldspar
(orthoclase) that resisted the deformation, dispersed through strongly foliated
aggregates of minerals that were deformed and recrystallized into smaller grains.
Through the centre of the photo runs a layer of intensely deformed rock (mylonite),
with a strong foliation and smaller residuals of feldspar. This rock would have
resembled the granite shown in Figure 72 before the deformation. Camera lens cap
for scale.

Figure 159.

Strongly heated, strongly folded, partly melted metamorphic rock (about 1,700 million years old) at Dingo Rock Hole, northeast of Alice Springs, central Australia. The light-coloured material (quartz and feldspar) is former melt that segregated into veins and solidified before the deformation. The deformation has been so intense that most of the light-coloured layers have been drawn out into thin, discontinuous lenses. The abundant remnants of small folds are evidence that the deformation was accomplished by flow.

Figure 160.

Even more intensely deformed rock (mylonite) than that shown in Figure 159, the
light-coloured layers having been drawn out into very small pods joined by extremely
thin lenses. Most of the folds have been obliterated by the intense deformation, but
the presence of a fold in a slightly less deformed part of the rock is evidence that the
deformation was due to flow. Papoose Flat, Inyo Mountains, California, United States.
The blue-and-white squares on the scale are one centimetre wide.

granite (Figures 3, 4, 70, 72) with strongly deformed granite (Figures 157, 158, 161, 162) and then with intensely deformed mylonite (Figures 157, 163).

Minerals may behave differently during intense deformation, some being stronger than others, though this varies with temperature. In fact, fracture and flow can occur simultaneously in different minerals in the same mylonite, as shown by fragments of strong feldspar or garnet surrounded by extremely elongated "ribbons" of recrystallized, weak quartz and mica

Figure 161.

This rock is similar to that shown in Figure 160, but has undergone a second episode of strong deformation, which folded the previously stretched feldspar grains and a narrow mylonite zone (like the one shown in Figure 160); Coles Point, Eyre Peninsula, South Australia. This is another spectacular example of multiple deformation of solid rocks by flow, during heating of rocks in Earth's crust. Camera lens cap for scale.

(Figures 30, 163). Some minerals (especially feldspar) break into fragments in lower temperature mylonite zones, but flow and recrystallize in higher temperature mylonite zones. Other minerals (for example, quartz and mica) tend to flow and recrystallize in all mylonite zones.

In many strongly deformed zones, the stretching is so intense that the stronger layers in the rock are broken, dismembered, and drawn out into isolated lenses (Figures 25, 159, 164).

Figure 162.

Microscope photo of a deformed granite (about 400 million years old), near Goulburn, New South Wales, Australia. Note the contrast between strong plagioclase crystals (large, fractured, shades of grey) and intensely recrystallized quartz (finely granular, grey, black and white) and drawn out aggregates of mica (bright colours). The plagioclase has resisted the deformation, whereas the quartz and mica have flowed in response to it. Interference colours; base of photo 2.4 centimetres.

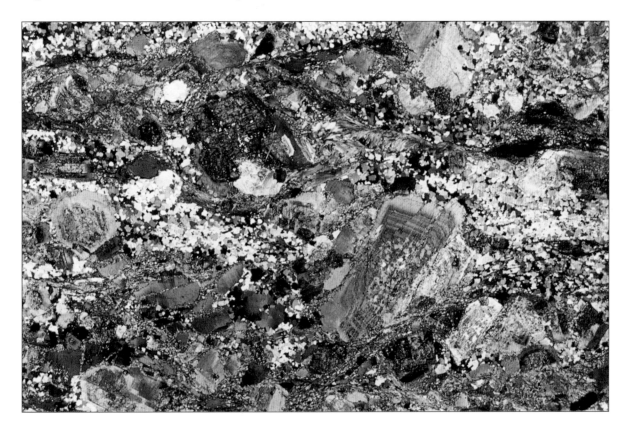

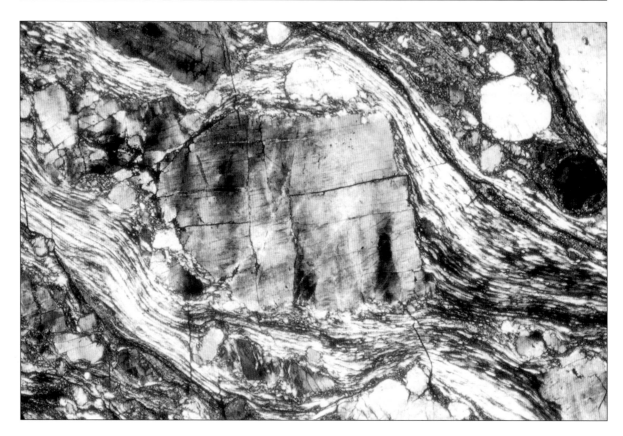

Figure 163.

Microscope photo of an intensely deformed mylonite (about 250
million years old) from east of Armidale, New South Wales, Australia.
Originally this was a granite like those shown in Figures 3 and 4. The
deformation was even more intense than that shown in Figure 162.
The photo shows a very marked contrast between strong plagioclase
crystals (large, fractured, shades of grey) and intensely recrystallized
(finely granular) and drawn out "ribbons" of quartz (shades of grey)
and biotite mica (brown). Though it has flowed a little and was then
fractured, the plagioclase grain in the centre of the photo was very
strong, and so tended to resist the deformation. In contrast, the
quartz and mica flowed around it and recrystallized to fine-grained
aggregates in response to the deformation. Interference colours; base
of photo 1.4 centimetres.

Figure 164.

Strongly deformed gneiss near Bettyhill, northwestern Scotland, showing deformation, breakage and separation of a layer of strong, dark rock. This layer was forced to break because it was not able to flow as easily as the rest of the rock. A group of students from the University of Leiden in The Netherlands is intently listening to a discussion of this structure – or maybe it's just so cold and windy that close to the cliff is the best place to stand!

From outer space

Identified flying objects

Earth is being constantly bombarded by pieces of rock from our solar system. Some come from the Moon and Mars, and some come from *comets,* which are small (1–10 kilometres across) bodies composed of rock fragments, dust, and ice that originate at the edges of the solar system. However, most are fragments of *asteroids* ("minor planets"), which are small metallic and rocky planet-like bodies. The largest is nearly 1,000 kilometres across. Three of them are more than 500 kilometres in diameter and 25 are more than 250 kilometres in diameter, but 85 percent are less than 100 kilometres in diameter. Asteroid particles range down to pebble-size objects, and even further down to cosmic dust particles a few microns across (though some cosmic dust also comes from outside the solar system).

Most asteroids are irregular in shape, and their surfaces are pitted with craters made by impacts with other asteroids, as shown in photographs taken from spacecraft (Figure 165). Asteroids mainly occur in a belt between Mars and Jupiter. More than 6,000 of them have been named. Though tens of thousands of asteroids exist in the solar system, they are widely separated. In fact, their total mass is only about 5 percent of the mass of the Moon.

Asteroids and fragments of asteroids that are on a collision course with Earth are called *meteoroids.* If they reach Earth's surface, we call them *meteorites.* About 700 meteoroids with

Figure 165.

The asteroid Ida (52 kilometres long) and its little moon Dactyl (1.5 kilometres across), which is seen as a small bright dot to the right of the asteroid. Note the very irregular shape of the asteroid. Although its normal colour is mostly grey, this photo was taken with a camera sensitive to near-infrared wavelengths of light, producing an enhanced colour image. The surface of Ida is covered by craters produced by repeated impacts of other asteroids. Their excellent state of preservation indicates that Ida is not internally active. The photo was taken 10,500 kilometres away by the spacecraft Galileo on August 28, 1993, on its way to Jupiter.

Photo courtesy of the National Space Science Data Center, through the World Data Center-A for Rockets and Satellites, the Galileo Project and Dr. Michael J. S. Belton.

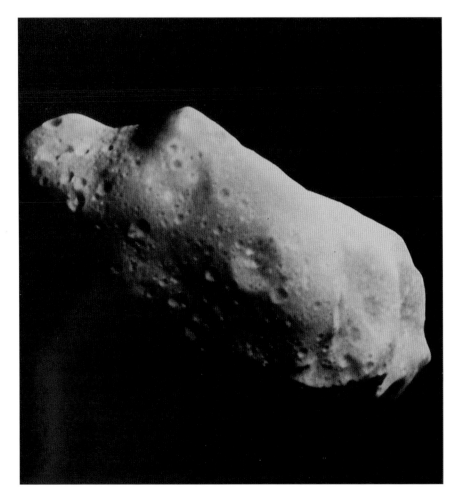

diameters of more than 1 kilometre are in an orbit that crosses Earth's path, but all are less than 30 kilometres in diameter.

Meteorites give us direct information on the nature of materials formed in the earlier stages of development of the solar system because many asteroids represent material that did not coalesce into larger planets when the solar system was formed about 4,600 million years ago.

In effect, the process of accretion of asteroid particles that formed Earth 4,600 million years ago is still occurring, though very slowly. Many thousands of meteorites, amounting to hundreds of tonnes of solid material, enter Earth's atmosphere every day, though most are melted and vapourized by friction with the atmosphere before reaching the planet's surface.

In addition, many small particles drift to the ground without being destroyed. Estimates of the amount of very small meteorites (less than a millimetre in diameter) are at least 4,000 tonnes per year; estimates of the amount of meteoroid dust particles (less than 0.06 millimetres in diameter and commonly less than 120 *micro*metres in diameter) range from a few thousand tonnes up to a million tonnes per year! Perhaps 30,000 tonnes per year is a fair estimate.

Meteoritic dust fragments (known formally as "interplanetary dust particles" and "cosmic spherules") have been discovered in stable ice sheets, deep sea sediments, impact sites on the surfaces of spacecraft, collector plates fitted to research aircraft sampling the stratosphere, and even swimming pool filters and roof drainpipes. Some particles are minute spheres of metal; others are non-metallic. Naturally, we don't notice the arrival of all this fine dust at Earth's surface, but we can see evidence of it in the faint glow in the sky just before dawn and after sunset, especially in the tropics, at the point where the sun is about to rise or has just set. The glow is caused by the scattering of sunlight by dust in interplanetary space of the inner solar system.

Much of it is presolar dust, which provides us with information on the materials and processes (such as dust accretion) in the solar system as it evolved 4,600 million years ago. In effect, interplanetary dust was the fundamental building block in the accretion process that produced the planets, asteroids, and comets. In addition, dust particles are being formed continually by the collision of asteroids with moons, Mars, and other asteroids.

It's very difficult to estimate the annual amount of larger meteorites that reach Earth. About 900 meteorites have been discovered in the last two hundred years. Recent estimates suggest that about 30,000 meteorites with masses of more than 100 grams reach Earth every year. Probably several hundred meteorites with a mass of more than a kilogram fall on Earth every year, but only about twenty of them are discovered. This is partly because most of them fall into the ocean, partly because many meteorites deteriorate quickly by weathering, and partly because the chances of witnessing a fall are very small. Only one of these falls occurs per million square kilometres annually on the average, which implies that the chances of a person being hit by a meteorite are very slight.

It has been estimated from astronomical observations of near-Earth asteroids that the chances of Earth being hit by a body one kilometre across are about 1 in 200,000–300,000 years. The chances of a collision of the type that probably killed the dinosaurs (see section on Meteorites and the Dinosaurs) is about 1 in several tens of millions of years. However, the chances of Earth being hit by one of the smaller 1,500 near-Earth asteroids a few hundreds of metres in diameter is about 1 in only 10,000 to 20,000 years. Several impact craters (see section on Impact Structures; Figures 166, 167) of medium size (several kilometres across) are younger than 5 million years, which suggests that very large impacts are certain to occur in the future – though it may be a very long while in human life terms.

Fortunately, most of the larger meteoroids don't reach us because they are swept away from Earth and the other inner planets by the strong gravity field of Jupiter. Many of the smaller meteorites are destroyed by the intense heat caused by friction with the air as they pass through the atmosphere at high speed (20–30 kilometres per second). Their surfaces melt, and the melted material is blown away. In contrast to larger meteorites, meteoritic dust particles are so small that they radiate heat very efficiently and so don't melt as they pass through the atmosphere.

We see evidence of meteorites travelling through Earth's atmosphere as bright streaks (*meteors*) we call "shooting stars" and "fireballs," though some of these are due to comets, as well as asteroid fragments. The brightness is due to the conversion of the meteorite to energy (some of which is released as light, as well as heat) as it is destroyed in the atmosphere. Many meteorites break up as they travel through the atmosphere, resulting in *meteorite showers.* Smaller meteorites slow down to less than the speed of sound as they pass through the atmosphere, but larger ones don't, and so cause "sonic booms."

Despite the extreme heating of the surfaces of larger meteorites as they pass through Earth's atmosphere, only the outer surface heats up. Their interiors remain cool because the duration of heating between their entry into the atmosphere and impact is only a few seconds, owing to their high speed. This enables glass to be preserved in some meteorites (see section on Meteorite Varieties), whereas it would crystallize if heated. The surface melting produces a dark "fusion crust" shown in Figure 168.

When the larger meteorites hit Earth, their huge amounts of kinetic energy (energy due to their mass and speed) are converted instantaneously into enormous amounts of heat, which can cause melting and even vapourization, not only of the meteorite itself, but also some of the impacted Earth rocks.

The largest meteorite found so far is the Hoba meteorite in Grootfontein, Namibia, South West Africa. It's so large that it's still in the ground, whereas most other meteorites found are now in museums. Its estimated mass is about 60 tonnes, but it has been extensively weathered since impact and so must have weighed more when it fell.

The best places to find meteorites are in dry areas without vegetation, such as dry lake beds (where the wind blows away dust and exposes the meteorites), flat plateaus of resistant rock, such as the Nullabor limestone surface in southwest Australia, and stable ice sheets, such as in Antarctica, where about 17,000 meteorite fragments have been found since 1969.

(text continues on page 195)

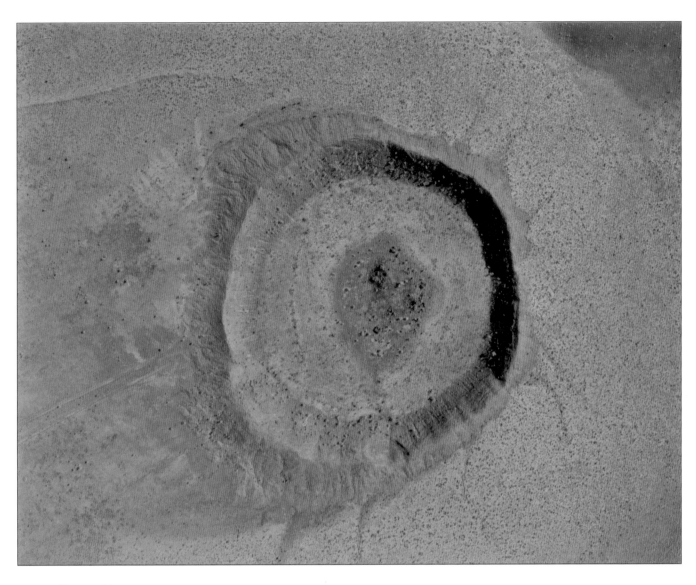

Figure 166.

Vertical air photo of the Wolfe Creek meteorite impact crater, 150 kilometres south of Halls Creek, Western Australia. The crater is 875 metres across. It has a bowl-shaped depression and a raised rim of rock debris (caused and ejected by the impact of a large meteorite about 300,000 years ago), which are typical of many impact craters. Oxidized iron meteorite fragments and glass formed by melting of the impacted rocks have also been found. Care must be taken to distinguish impact craters from collapse structures (calderas) formed by gigantic explosive volcanic eruptions, such as the one shown in Figure 93.

Photo courtesy of Kevron Aerial Surveys Pty. Ltd., Perth, Western Australia.

Figure 167.

The well-preserved Meteor Crater (Barringer Crater)
near Flagstaff, Arizona, United States, which is 1.2
kilometres in diameter and over 200 metres deep, was
formed about 49,000 years ago by the impact of a
large iron meteorite. The crater has a bowl-shaped
central depression and a raised rim, which is typical of
many impact craters. Though many small meteorite
fragments have been found here, most were
vapourized by the immense force of the impact.
Photo by Scott Johnson.

Figure 168.

Sawn slab, 7 centimetres long, of a stony meteorite (the Barratta chondrite)
reported as having been found in New South Wales, Australia, in 1845. The
chondrite contains abundant, well-defined chondrules, and has a dark skin ("fusion
crust"). The slab gives a clear impression of the formation of the meteorite by the
aggregation of chondrules and other asteroid fragments.

Photo courtesy of Alex Bevan,
Western Australian Museum, Perth.

Meteorites found in Antarctica are preserved in ice and so are much less weathered and con-
taminated than those found elsewhere. Metal detectors and magnets can be used to help find
meteorites containing iron.

Impact structures

Obvious evidence of large meteorite or asteroid impacts, apart from discoveries of meteorite fragments themselves, are circular craters with raised rims, surrounded by rock fragments thrown out of the crater (Figures 166, 167). Many of these fragments contain characteristic minerals formed when the rocks were subjected to extremely high pressures for
short periods.

Impact structures are rare on Earth, owing to erosion processes that tend to remove the
evidence of craters with time, especially in areas of high rainfall. However, some spectacular
craters have been preserved in deserts (Figures 166, 167).

Only about 160 confirmed impact structures have been recognized, and another three or four are discovered each year. Their diameters range from less than 1 to more than 200 kilometres. The diameters of craters are much larger than the impacting meteorite, and the depth is about one-tenth of the crater diameter. Most impact structures on Earth are younger than about 2,000 million years old. In fact, most preserved craters are younger than 200 million years old because older impact structures on Earth have been obliterated by erosion or buried by younger sedimentary rocks.

This scarcity of impact craters on Earth is in marked contrast to the surfaces of the Moon, Mars, and Mercury, which show superimposed craters everywhere, some being up to 900 kilometres in diameter. These craters mainly reflect a massive bombardment by very large meteorites, which is believed to have ceased about 3,800 million years ago, and that also occurred on Earth. The lack of water and therefore the absence of erosion, as well as the absence of the crustal movement that characterizes Earth (Chapter 2), have preserved impact craters on the Moon and Mars, in contrast to the surface of Earth.

The heavy bombardment of Earth by very large meteorites 3,800–4,200 million years ago involved a rate of meteorite arrivals a hundred times higher than the rate since that time. The early bombardment may have been so intense that it shattered and melted the rocks outcropping on Earth's surface. Some geologists have even suggested that the bombardment may have inhibited the development of life until it subsided to a level similar to that of the present day.

Meteorites and the dinosaurs

A massive meteorite impact on Earth occurred 65 million years ago. It threw up a huge amount of dust and other debris into the atmosphere, blocking out the Sun's energy for perhaps six months in some places, and reducing the air temperature very rapidly. At least half of the animal species on Earth, including the mighty dinosaurs, and many plants did not survive.

The main evidence of this huge impact is a thin clay layer that is richer than usual in the chemical element iridium (though it is only present in minute amounts all the same). This layer has been discovered sandwiched between other sedimentary rocks in many places on Earth's surface. Iridium is rare in common Earth rocks, but much more abundant in meteorites, making it a reliable indicator of debris showered from a meteorite impact.

But where is the impact crater corresponding to this devastating event? The most likely structure is a very large, multi-ringed basin, 170 kilometres across, that is buried beneath several hundred metres of sediment, around the village of Chicxulub in the Yucatan Peninsula of Mexico. It has the right age and size, and has been revealed by geophysical techniques. An asteroid, 8–9 kilometres across, is believed to have been responsible for this impact structure.

Similar massive episodes of extinction of many of Earth's animal and plant species have occurred at other times in the distant past. They may have been caused by giant meteorite impacts, but explosive volcanic activity on an unusually grand scale could have had a similar effect; a large drop in temperature could have been caused by discharge of huge amounts of volcanic fragmental material into the atmosphere. It is also possible that such volcanic activity was caused by giant meteorite impacts.

Meteorite varieties

Meteorites come in three main broad types: *"stony meteorites"* (Figures 168–171), *"iron meteorites"* (Figures 172, 173) and *"stony iron meteorites"* (Figure 174), though meteorites are very variable. Falls of stony meteorites are much more common than falls of iron meteorites, but iron meteorites are easier to find. This is because iron meteorites are much more distinctive, being unlike common Earth rocks, and because stony meteorites weather and disintegrate more quickly. Of the meteorites examined so far, about 94 percent are stony meteorites, about 5 percent are iron meteorites, and the rest are stony iron meteorites.

The "stones" are the most variable of meteorites, and consist mainly of silicates of magnesium and iron (mainly olivine and pyroxene – minerals common in Earth's mantle) with some (less than 25 percent) iron-nickel metal. Some examples contain carbon, sulphur, and water, mostly combined in various minerals.

Many stony meteorites, known as *chondrites,* contain round globules ("chondrules") containing spiky or branch-like crystals, the shapes of which suggest that the globules were once droplets of melted rock (Figures 169–171). The droplets originally may have been dustballs that were melted by flash-heating events in the solar nebula, before accretion of the globules into asteroids. The cause of the heating is unknown, but may be shock waves. Chondrites were formed by the accretion of a wide variety of rocks with very different pre-accretion histories in the solar nebula. Stony meteorites without chondrules are known as *achondrites.*

The "irons" consist largely or entirely of metal, which is mainly iron with some nickel (up to 30 percent, generally 5–20 percent) and smaller amounts of other metals. In some

(text continues on page 202)

Figure 169.

Microscope photo of a stony meteorite (chondrite) from Wanaaring, New South Wales, Australia, showing spherical globules ("chondrules") rich in branching to spiky crystals, mainly of pyroxene, interspersed with granular olivine and pyroxene and patches of iron-nickel metal (black, because it doesn't transmit light). Normal colours; base of photo 1.75 millimetres.

Specimen courtesy of The Australian Museum, Sydney.

Figure 170.

Microscope photo of a stony meteorite (chondrite) from east of Tillibigeal, New South Wales, Australia, showing a spherical globule ("chondrule") formed by the very rapid cooling of a melt droplet. The chondrule consists of olivine (bright colours) in glass (black). The olivine crystals have skeletal shapes similar to those formed in rapidly cooled basalts (Figure 65). Interference colours; base of photo 1.75 millimetres.

Thin section courtesy of The Australian Museum, Sydney.

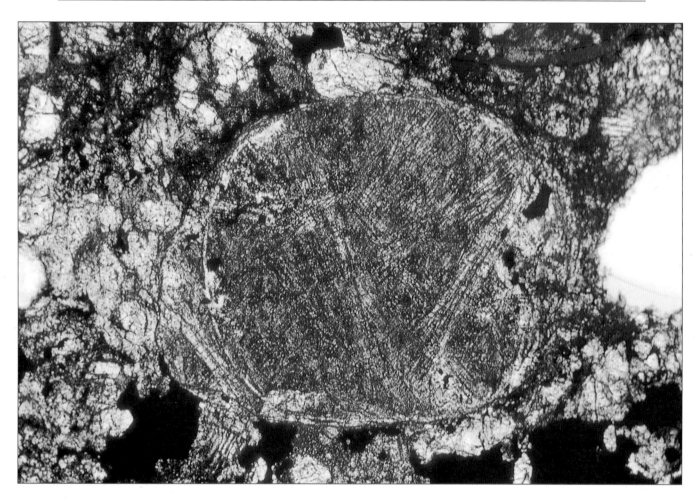

Figure 171.

Microscope photo of a stony meteorite (chondrite)
from Wanaaring, New South Wales, Australia, showing
a spherical globule ("chondrule") consisting of very
spiky pyroxene crystals similar to those occurring in
rapidly cooled melts (Figures 64, 65, 101). Normal
colours; base of photo 1.75 millimetres.

Thin section courtesy of The Australian Museum, Sydney.

Figure 172.

Iron meteorite from Mundrabilla, Nullabor Plain, Western Australia. The meteorite has a rounded, pitted surface, caused mainly by uneven melting due to extreme heating of the iron as it passed though Earth's atmosphere. The iron would be bright and metallic (as shown in Figure 173) underneath the dull, oxidized surface. The meteorite weighs 600 kilograms.

Specimen on display at The Australian Museum, Sydney.

Figure 173.

Iron meteorite from Namibia, which has been sawn and etched with nitric acid to
reveal the characteristic crosshatch pattern ("Widmanstätten structure") formed by
nickel-rich lamellae with iron-rich metal in between. The parallelism of the sets of
lamellae across the whole specimen shows that this is part of a single crystal (in
which the lamellae have segregated during slow cooling). The extremely large size of
the crystal is consistent with it having grown and cooled very slowly in the hot core
of a planetary body (asteroid). The coin is 2.8 centimetres in diameter.

Specimen courtesy of The Australian Museum, Sydney.

varieties, iron-rich and nickel-rich metal alloys tend to segregate into separate lamellae,
forming a spectacular criss-cross pattern (Figure 173). This segregation occurred during slow
cooling of the metal in the parent asteroid. Metallic iron typically doesn't occur naturally on
Earth's surface (although it does occur, alloyed with nickel, as small grains in a few rocks),
because it rusts so quickly in the atmosphere. Therefore, when we find naturally occurring
metallic iron, we know it is most probably of extraterrestrial origin. The nickel slows down
the rusting of the iron, which helps preserve iron meteorites on Earth's surface.

Figure 174.

Sawn and polished slab of a stony iron meteorite from
Finmarken, Norway, showing shiny metallic iron
interspersed with olive-green to dull brown, weathered
olivine and pyroxene. The coin is 2.8 centimetres in
diameter.

Specimen courtesy of The Australian Museum, Sydney.

The "stony irons," as the name suggests, contain large amounts of both iron-nickel metal
(averaging around 50 percent) and magnesium-iron silicate (olivine and pyroxene), as
shown in Figure 174.

A more genetic classification separates meteorites into "primitive" meteorites, which
were not processed into a planet, and "evolved" meteorites, which have been through a major
planet-forming process similar to that responsible for the various shells in Earth (Figure 8).
Chondrites are primitive, and achondrites, stony irons, and irons are evolved, as discussed in

more detail below. The primitive nature of chondrites makes them important for deciphering processes that occurred in the earliest stages of the formation of the solar system.

How old are meteorites?

Radioactive dating of primitive meteorites (chondrites) reveals a consistent age of about 4,600 million years, which is inferred to be the age of Moon rocks. Because neither primitive meteorites nor Moon rocks have been altered chemically since their formation (in contrast to Earth rocks, which are constantly being changed and chemically recycled), this is generally taken to be the age of formation of the solar system, including Earth. Therefore, primitive meteorites probably represent materials condensed from gases in the solar nebula, accreted to form solid bodies at the same time as Earth, and then repeatedly re-fragmented. If so, they may well be the debris left over from the materials that coalesced to form the nine planets and the thousands of asteroids. This idea is supported by the fact that the chemical composition of chondrites is very similar to that of the Sun, less the gaseous elements hydrogen and helium.

As mentioned in Chapter 2, Earth's core is mainly iron metal, whereas the mantle consists mainly of minerals such as olivine and pyroxene, which mainly contain magnesium, silicon and oxygen, with calcium, iron and aluminium. This contrast matches the difference between iron and stony meteorites, and so confirms, in a general way, that many meteorites are derived from disintegrated larger planetary bodies somewhat similar to Earth.

Therefore, iron meteorites are samples of cores of asteroids that have evolved and "differentiated" into shells like those of Earth, after formation of the asteroids by accretion of fragments in the solar nebula. Whereas stony iron meteorites come from the core-mantle boundary regions of these differentiated asteroids, achondrites are mostly fragments of igneous rocks (Chapter 5) formed at or near the surface of differentiated asteroids. In contrast, chondrites come from parent asteroids that escaped post-accretion differentiation. More than eighty known asteroids are of chondrite composition.

As mentioned previously, spacecraft photographs show impact craters on the surfaces of asteroids (Figure 165), indicating bombardment by other asteroids. Evidence of bombardment is also provided by meteorites showing mineral deformation and especially by meteorites with fragmental structures (breccias).

Epilogue

Seen one rock . . . ?

I hope that I have convinced you that Earth rocks are as fascinating in their variety as they are beautiful. If you've seen one rock, you certainly haven't seen them all! After forty-five years of looking at rocks, I am still amazed at the new varieties I see.

Museums are good places to learn more about Earth's rocks and minerals, as well as the marvellous animals and plants that lived on the planet in the distant past. The displays in museums are spectacular as well as informative, and the people who look after the collections and displays are dedicated and helpful.

I recommend the following books to those who would like to read more deeply about the rocks of Planet Earth, as well as something on the solar system and the universe.

Busbey, A. B., Coenraads, R. R., Roots, D., & Willis, P. 1997. *The Nature Company Guides – Rocks and Fossils.* RD Press, in association with The Nature Company Guides. (A well-illustrated introductory book on minerals, rocks, and fossils).

Dickey, J. S. 1996. *On the Rocks. Earth Science for Everyone.* Wiley, New York. (An engaging book full of interesting discussions of a wide range of topics, from atoms and crystals to the solar system).

Flaum, E. 1988. *The Planets. A Journey Into Space.* Crescent Books, New York. (An excellently illustrated book, full of the wonder and excitement of astronomy).

Henbest, N. (with photos by P. Chudy) 1992. *The Universe. A Voyage Through Space and Time.* Wiedenfeld & Nicolson, London. (A spectacularly illustrated book with clear discussions about the history of the universe and the nature of stars, planets, and moons).

Plimer, I. R. 1997. *A Journey Through Stone.* Reed Books, Australia. (A fascinating account of the geological history, minerals, and mining of the Chillagoe area, north Queensland, explaining Earth processes in a simple way).

Skinner, B. J. & Porter, S. C. 1995. *The Blue Planet. An Introduction to Earth System Science.* Wiley, New York. (A wonderful, well-illustrated book about all aspects of Planet Earth, taking you to first-year university level).

Glossary

Mineral names

amphibole: A group of water-bearing silicate and aluminosilicate minerals with variable amounts of calcium, sodium, magnesium, iron, and titanium; formed in igneous and metamorphic rocks.

andalusite: aluminosilicate; formed at moderately high temperatures in metamorphic rocks.

calcite: calcium carbonate; common in limestones (for example, coral reefs) and fossil shells; soft (scratches easily).

chalcopyrite: copper iron sulphide; has yellowish brassy metallic lustre, commonly with a coppery tarnish formed by weathering.

chlorite: water-bearing silicate with magnesium, iron and aluminium; generally dark green, soft shiny flakes; formed at relatively low temperatures.

chloritoid: aluminosilicate of iron, magnesium, with some water, occurring in metamorphosed shales.

chrysotile: variety of serpentine (q.v.) occurring in veins; one of the main types of asbestos; soft, commonly with a greasy feel.

clay minerals: a group of water-rich aluminosilicate minerals with various other elements, mainly potassium, magnesium, and iron; formed at low temperatures, especially during weathering of rocks in Earth's atmosphere.

covellite: copper sulphide; bright blue colour.

epidote: calcium-iron aluminosilicate with water; generally pistachio green.

feldspars: aluminosilicates of sodium, calcium, and potassium; the most common group of minerals, present in most rocks in Earth's crust; the main members of the group are **plagioclase** (sodium-calcium feldspar) and **orthoclase** (potassium feldspar).

galena: lead sulphide; grey with a metallic lustre.

garnet: aluminosilicate of very variable chemical composition, but mainly with iron, magnesium, manganese, and/or calcium; commonly dark brownish to pinkish red; semi-precious gemstone.

hematite: red iron oxide; not magnetic.

ice: mineral (if it occurs naturally) with the chemical composition of water.

magnetite: black iron oxide and magnetic, as the name suggests; common in Earth's crust.

mica: group of aluminosilicate minerals with potassium and water; soft, shiny and with a marked sheet-like structure; the main micas are **biotite** (dark coloured, rich in iron and magnesium) and **muscovite** (light coloured, rich in aluminium).

olivine: silicate of magnesium and iron; abundant in Earth's mantle and in stony meteorites; olive-green; the gem variety (shiny olive-green with emerald-coloured internal reflections) is **peridot.**

opal: a non-crystalline variety of silicon dioxide with some water, which is commonly milky in appearance, but may show spectacular colours in the gem variety.

orthoclase: see feldspar.

plagioclase: see feldspar

potassium feldspar: see feldspar

prehnite: aluminosilicate of calcium with water; formed at relatively low temperatures.

pumpellyite: aluminosilicate of magnesium, iron and calcium with water; formed at relatively low temperatures.

pyrite: iron sulphide; brassy metallic lustre.

pyroxene: silicate of iron, magnesium and calcium, with variable composition; some varieties contain sodium, aluminium, or titanium.

quartz: silicon dioxide; hard, with vitreous (glassy) lustre.

serpentine: water-rich silicate of magnesium; soft, commonly with greasy feel; formed at low temperatures.

sillimanite: aluminosilicate; formed at high temperatures in metamorphic rocks.

sphalerite: zinc sulphide; generally dark grey and lustrous.

staurolite: aluminosilicate of iron, with some water, and small amounts of other elements; red-brown crystals; occurring in metamorphic rocks.

sulphide minerals: metal sulphides, such as pyrite (iron sulphide), galena (lead sulphide), chalcopyrite (copper-iron sulphide), and sphalerite (zinc sulphide); generally shiny and metallic in appearance.

zeolites: a group of water-bearing aluminosilicate minerals with sodium and/or calcium; formed at relatively low temperatures; generally fibrous and light coloured.

Rock names

basalt: fine-grained, dark grey to black, igneous rock rich in plagioclase and pyroxene, with or without olivine; crystallized on or just below Earth's surface.

breccia: rock composed of large, angular fragments; formed in faults (**fault breccia**) or volcanic eruptions (**volcanic breccia**). Meteorite breccias are derived from asteroids that have undergone impacts from other asteroids.

conglomerate: sedimentary rock composed of large rounded pebbles.

fault breccia: see breccia.

gabbro: coarse-grained, dark grey igneous rock rich in plagioclase and pyroxene, with or without olivine; crystallized below Earth's surface.

glass (volcanic glass): rock formed by the very rapid cooling (freezing) of lava in the atmosphere or water, the cooling being too fast for crystals to develop; has the rigidity of a solid, but the atomic arrangement of a melt. Glass also occurs in spherical droplets (chondrules) in chondritic meteorites.

gneiss: coarse-grained, high-temperature metamorphic rock, generally with a relatively weak foliation.

granite: coarse-grained igneous rock rich in quartz and feldspar, with or without mica or amphibole; crystallized below Earth's surface.

ice (glacier ice): rock consisting entirely of the mineral, ice.

kimberlite: fragmental rock (breccia) from deep in Earth's mantle; may carry diamonds.

limestone: rock composed mainly of calcium carbonate (calcite).

migmatite: rock formed by partial melting during high-temperature metamorphism, consisting partly of light-coloured melt (now solidified) and partly of dark-coloured unmelted rock.

mylonite: fine-grained, strongly foliated rock formed by intense deformation in relatively local high-strain zones in Earth's crust.

pegmatite: very coarse-grained igneous rock, typically rich in quartz and feldspar.

peridotite: rock rich in olivine; characteristic rock of Earth's mantle.

pumice: volcanic glass with many bubble shapes ("glass froth").

rhyolite: fine-grained, generally light-coloured igneous rock rich in quartz and feldspar, with or without mica or amphibole; crystallized on or just below Earth's surface.

sandstone: rock formed by the accumulation of sand size fragments (2 millimetres to $\frac{1}{16}$ of a millimetre in diameter) cemented together.

schist: strongly foliated metamorphic rock, rich in mica, generally with other minerals such as andalusite, garnet, or staurolite; formed by the metamorphism of shale.

shale: fine-grained sedimentary rock composed of very small, clay size particles (smaller than $\frac{1}{256}$ of a millimetre in diameter); very rich in aluminium, owing to an abundance of clay minerals.

slate: fine-grained metamorphic rock with a strong foliation ("slaty cleavage") formed by the deformation and weak heating of shale; used as roofing and floor tiles because of the ease of splitting parallel to the cleavage.

volcanic breccia: see breccia

Other terms

aa: jagged, clinkery basalt formed by fragmentation of a lava flow that has become too viscous to flow easily; Hawaiian word.

ash: fragmental material of sand size ejected from a volcano; consists of fragments of crystals and volcanic glass.

crystal: mineral with regular crystal faces (as shown by many mineral specimens in museums).

glass: see volcanic glass

grain: mineral without crystal faces.

lava: molten rock plus crystals that is capable of flowing, ejected from a volcano.

magma: molten rock plus crystals that is capable of flowing, either on or beneath Earth's surface

pahoehoe: basalt with a ropy surface structure caused by flow; Hawaiian word.

volcanic ash: (see ash)

volcanic glass: (see glass)

Index